Big Data in Materials Research and Development

SUMMARY OF A WORKSHOP

Maureen Mellody, *Rapporteur*

Defense Materials Manufacturing and Infrastructure Standing Committee

Division on Engineering and Physical Sciences

NATIONAL RESEARCH COUNCIL
OF THE NATIONAL ACADEMIES

THE NATIONAL ACADEMIES PRESS
Washington, D.C.
www.nap.edu

THE NATIONAL ACADEMIES PRESS   500 Fifth Street, NW   Washington, DC 20001

NOTICE: The project that is the subject of this report was approved by the Governing Board of the National Research Council, whose members are drawn from the councils of the National Academy of Sciences, the National Academy of Engineering, and the Institute of Medicine.

This study was supported by Contract/Grant No. W911NF-11-C-0212 between the National Academy of Sciences and the Department of Defense. Any opinions, findings, or conclusions expressed in this publication are those of the author(s) and do not necessarily reflect the views of the organizations or agencies that provided support for the project.

International Standard Book Number-13: 978-0-309-30379-8
International Standard Book Number-10: 0-309-30379-6

Cover image: Big Data is partially driven by today's ever increasing ability to electronically generate and store data. Big Data is also requiring ever increasing computational power to fully mine and understand the data. Thus, beside Big Data itself, at the core of this topic is today's materials research and development enabling the bootstrapping of the complex electronic materials needed for the computations. Artist: Erik Svedberg. Image created electronically and algorithmically.

This report is available in limited quantities from

National Materials and Manufacturing Board
500 Fifth Street, NW
Washington, DC 20001
nmmb@nas.edu
http://www.nationalacademies.edu/nmmb

Additional copies of this report are available for sale from the National Academies Press, 500 Fifth Street, NW, Keck 360, Washington, DC 20001; (800) 624-6242 or (202) 334-3313; http://www.nap.edu/.

Copyright 2014 by the National Academy of Sciences. All rights reserved.

Printed in the United States of America

THE NATIONAL ACADEMIES
Advisers to the Nation on Science, Engineering, and Medicine

The **National Academy of Sciences** is a private, nonprofit, self-perpetuating society of distinguished scholars engaged in scientific and engineering research, dedicated to the furtherance of science and technology and to their use for the general welfare. Upon the authority of the charter granted to it by the Congress in 1863, the Academy has a mandate that requires it to advise the federal government on scientific and technical matters. Dr. Ralph J. Cicerone is president of the National Academy of Sciences.

The **National Academy of Engineering** was established in 1964, under the charter of the National Academy of Sciences, as a parallel organization of outstanding engineers. It is autonomous in its administration and in the selection of its members, sharing with the National Academy of Sciences the responsibility for advising the federal government. The National Academy of Engineering also sponsors engineering programs aimed at meeting national needs, encourages education and research, and recognizes the superior achievements of engineers. Dr. C. D. Mote, Jr., is president of the National Academy of Engineering.

The **Institute of Medicine** was established in 1970 by the National Academy of Sciences to secure the services of eminent members of appropriate professions in the examination of policy matters pertaining to the health of the public. The Institute acts under the responsibility given to the National Academy of Sciences by its congressional charter to be an adviser to the federal government and, upon its own initiative, to identify issues of medical care, research, and education. Dr. Victor J. Dzau is president of the Institute of Medicine.

The **National Research Council** was organized by the National Academy of Sciences in 1916 to associate the broad community of science and technology with the Academy's purposes of furthering knowledge and advising the federal government. Functioning in accordance with general policies determined by the Academy, the Council has become the principal operating agency of both the National Academy of Sciences and the National Academy of Engineering in providing services to the government, the public, and the scientific and engineering communities. The Council is administered jointly by both Academies and the Institute of Medicine. Dr. Ralph J. Cicerone and Dr. C. D. Mote, Jr., are chair and vice chair, respectively, of the National Research Council.

www.national-academies.org

PLANNING COMMITTEE FOR WORKSHOP ON BIG DATA IN MATERIALS RESEARCH AND DEVELOPMENT

MICHAEL F. McGRATH, Analytic Services, Inc., *Chair*
VALERIE BROWNING, ValTech Solutions, LLC
JESUS M. DE LA GARZA, Virginia Polytechnic Institute and State University (Virginia Tech)
ROSARIO GERHARDT, Georgia Institute of Technology
PAUL KERN, U.S. Army (retired)
ROBERT H. LATIFF, R. Latiff Associates
E. WARD PLUMMER, Louisiana State University
ROBERT E. SCHAFRIK, GE Aviation
DENISE F. SWINK, Independent Consultant
HAYDN N.G. WADLEY, University of Virginia

Staff

JAMES LANCASTER, Acting Director
ERIK B. SVEDBERG, Senior Program Officer
MAUREEN MELLODY, Rapporteur
JOSEPH PALMER, Senior Project Assistant

DEFENSE MATERIALS MANUFACTURING AND INFRASTRUCTURE STANDING COMMITTEE

MICHAEL F. McGRATH, Analytic Services, Inc., *Chair*
VALERIE BROWNING, ValTech Solutions, LLC
JESUS M. DE LA GARZA, Virginia Tech
ROSARIO GERHARDT, Georgia Institute of Technology
PAUL KERN, U.S. Army (retired)
ROBERT H. LATIFF, R. Latiff Associates
E. WARD PLUMMER, Louisiana State University
ROBERT E. SCHAFRIK, GE Aviation
DENISE F. SWINK, Independent Consultant
HAYDN N.G. WADLEY, University of Virginia

Staff

JAMES LANCASTER, Acting Director
ERIK B. SVEDBERG, Senior Program Officer
HEATHER LOZOWSKI, Financial Associate
JOSEPH PALMER, Senior Project Assistant

Acknowledgment of Reviewers

This report has been reviewed in draft form by individuals chosen for their diverse perspectives and technical expertise, in accordance with procedures approved by the National Research Council's Report Review Committee. The purpose of this independent review is to provide candid and critical comments that will assist the institution in making its published report as sound as possible and to ensure that the report meets institutional standards for objectivity, evidence, and responsiveness to the study charge. The review comments and draft manuscript remain confidential to protect the integrity of the deliberative process. We wish to thank the following individuals for their review of this report:

David Aspnes (NAS), North Carolina State University,
Valerie Browning, ValTech Solutions, LLC,
Jesus de la Garza, Virginia Tech, and
Sylvia Johnson, NASA Ames Research Center.

Although the reviewers listed above have provided many constructive comments and suggestions, they were not asked to endorse the views presented at the workshop, nor did they see the final draft of the workshop summary before its release. The review of this workshop summary was overseen by David W. Johnson, *Journal of the American Ceramic Society*. Appointed by the NRC, he was responsible for making certain that an independent examination of this workshop summary was carried out in accordance with institutional procedures and that all review comments were carefully considered. Responsibility for the final content of this summary rests entirely with the author and the institution.

Contents

OVERVIEW   1

WORKSHOP THEMES   3
    Data Availability, 3
    Data Size: "Big Data" vs. Data, 6
    Quality and Veracity of Data and Models, 7
    Data and Metadata Ontology and Formats, 8
    Metadata and Model Availability, 8
    Culture, 9

SESSION 1: INTRODUCTION TO BIG DATA   11
    Frontiers in Massive Data Analysis and Their Implementation, 11
    IBM and Big Data, 16
    Big Data for Biosecurity, 18
    Discussion, 21

SESSION 2: BIG DATA ISSUES IN MATERIALS RESEARCH AND   23
DEVELOPMENT
    Physics in Big Data, 23
    Materials Genome Initiative and Big Data, 26
    General Electric Efforts in Materials Data: Development of the
       ICME-Net, 31

Smart Manufacturing: Enterprise Right Time, Networked Data,
    Information, and Action, 34
Discussion, 39

SESSION 3: BIG DATA ISSUES IN MANUFACTURING   41
    Data Needs to Support ICME Development in DARPA Open
        Manufacturing, 41
    The Materials Information System, 44
    Discussion, 49

SESSION 4: THE WAY AHEAD   50
    Lightweight and Modern Metals Manufacturing Innovation Institute:
        Implications for Materials, Manufacturing, and Data, 50
    Direction of Policy, 54

REFERENCES   57

APPENDIXES

A	Workshop Statement of Task	61
B	Workshop Participants	62
C	Workshop Agenda	64
D	Acronyms	66

Overview

The Defense Materials Manufacturing and Infrastructure (DMMI) Standing Committee convened a workshop on February 5 and 6, 2014, to discuss the impact of big data on materials and manufacturing. The DMMI standing committee is organized under the auspices of the National Materials and Manufacturing Board of the National Research Council (NRC) and with the sponsorship of Reliance 21, a Department of Defense (DOD) group of professionals that was established in the DOD science and technology community to increase awareness of DOD science and technology activities and increase coordination among the DOD services, components, and agencies.

This report has been prepared by the workshop rapporteur as a factual summary of what occurred at the workshop. The planning committee's role was limited to planning and convening the workshop. The views contained in the report are those of individual workshop participants and do not necessarily represent the views of the workshop participants as a whole, the planning committee, or the National Research Council.

To conduct its workshop on big data, the DMMI standing committee first organized a workshop planning committee to identify workshop topics and agenda items, speakers, and guests to be invited. The planning committee consulted with Reliance 21 and members of the community to develop and organize the workshop. The workshop was held at the National Academy of Sciences, 2101 Constitution Ave., NW, Washington, D.C. Approximately 50 participants, including speakers, members of the DMMI standing committee, Reliance 21, invited guests, and members of the public, participated in the two-day workshop.

Some of the topics addressed at the workshop included these:

- Any unusual aspects of materials and manufacturing needs compared to other big data efforts.
- Data ownership and access, including materials property data.
- Collaboration and the exploitation of big data's capabilities, including information exchange, validation, and security.
- Cost and ease of maintenance of data, including any associated infrastructure needed.
- Assuring the pedigree of the data and metadata.
- A general understanding of the materials community's wants and needs in big data.

To assist the reader, a short summary of recurring themes from the workshop presentations and discussions is presented below. These six themes are merely a short description of the items that were discussed by multiple speakers or participants during the course of the workshop. They were identified for this report by the rapporteur, not by the workshop participants.

1. Data availability.
2. Data size: "big data" vs. data.
3. Quality and veracity of data and models.
4. Data and metadata ontology and formats.
5. Metadata and model availability.
6. Culture.

Within each theme, the discussion addressed current concerns, needs, or requirements—in other words, "challenges"—and potential improvements.

After briefly describing the recurring themes, the report summarizes the workshop presentations and discussions. Appendix A contains the statement of task for the workshop. Appendix B lists the workshop participants. Appendix C is the workshop agenda, and Appendix D defines acronyms used in this report.

Workshop Themes

Much of the workshop discussion was driven by an overarching assumption: The materials science community would benefit from appropriate access to data and metadata for materials development, processing, application development, and application life cycles. Currently, that access does not appear to be sufficiently widespread, and many participants captured the constraints and identified potential improvements to enable broader access to materials and manufacturing data and metadata.

DATA AVAILABILITY

Data availability was a much discussed topic at the workshop. Several participants spoke of the difficulties associated with the fact that data are not always archived properly for long-term storage and that experimental data, including information about the procedures used and how the data were acquired, are not readily available in the public domain. Among the obstacles to data sharing discussed at the workshop were these:

- *Intellectual property constraints.* Several participants said that data ownership is an obstacle to data sharing. Thom Mason, of Oak Ridge National Laboratory, said that because individual researchers in the materials research community invest substantial time and effort in making and characterizing a sample, there will be challenges associated with moving into open source data. Other participants noted that access to proprietary and other data owned by private industry can also be constrained because a company wishes to retain its own intellectual property. This then becomes a cultural consideration in the community (see the final theme in this chapter: Culture).

In discussion sessions, some participants also identified several challenges when the data are generated under a DOD contract. DOD contracts have many data requirements in them, and DOD needs to be aware that these requirements can sometimes be perceived as cost prohibitive. One participant pointed out that material suppliers, original equipment manufacturers, and the government have a joint responsibility to share data. In some cases, a participant noted, suppliers provide the material but no corresponding metadata. Agreements with suppliers can take a year or two to develop, which can slow the rate of innovation. Julie Christodoulou, of the Office of Naval Research (ONR), stated that the Lightweight and Modern Metals Manufacturing Innovation Institute (LM3II) is currently sorting through issues related to levels of engagement and intellectual property. She believes that each project will likely have its own unique intellectual property arrangements. Other participants indicated that the Air Force Office of Scientific Research (AFOSR), ONR, the Defense Advanced Research Projects Agency (DARPA), and other defense agencies are exploring ways to share with the broader community data generated under a DOD contract.

- *Data heterogeneity.* Several participants pointed out that data are produced by many different people and organizations. They noted that when data are produced via heterogeneous distributed systems, there is no easy or centralized access.

Several possible remedies for the unavailability of data were explored at the workshop:

- *Data-mining utilities.* A participant noted the possibility of using data-mining utilities—something akin to a "materials Google"—to search existing data. This could improve access and increase ease of use.
- *Storing materials instead of data.* On several occasions, participants discussed the option of storing samples of a material directly instead of storing data related to an experiment. During a discussion session, Denise Swink, a private consultant, asked if reestablishing critical material repositories in DOD should be considered. She said there is a reluctance to do so but noted the large financial difference between stockpiling and archiving—a few of the workshop participants are probably more interested in archiving. One participant mentioned the Air Force's digital twin program, in which the digital representation of a material keeps informa-

tion about the material properties. Perhaps it would be valuable to also save an actual sample to examine along with the digital twin. However, some workshop participants pointed out that critical questions would need to be addressed, such as how much material to retain, access criteria, and other policy issues. Others suggested that any sort of repository could be prohibitively expensive. To be a viable option, it might be necessary, they suggested, to set priorities for which materials to retain.

- *Government mandate to store data.* Several participants suggested that a data storage mandate might be useful, proposing that any government-funded data be put into a standard format and stored in some long-term external repository. A participant reported that this step is under consideration by the National Science Foundation (NSF) and the Department of Energy (DOE). A DOD participant said that DOD had explored the idea of such a mandate but found it to be time-consuming and expensive and, in the end, not cost-effective. He said that a mandate would have to be for more than a data format; it would need to answer questions such as who owns and maintains the data and where the information should reside. Someone else noted that NSF has begun developing a data repository that is not, however, very user friendly.

- *Development of private data repositories and formats.* Workshop participants also discussed data repositories and how to ensure that data remain accessible in the future. Several persons spoke of the need to define structured databases. Dan Crichton, of NASA's Jet Propulsion Laboratory (JPL), said that data archives typically require at least a 50-year expected time window for operability. For data to remain usable for that long, he pointed out, they would need to be captured in a stable format, even if that format is not contemporary, and should be software independent, relying upon static conceptual models. He noted in particular the importance of separating the data from the technology used to acquire it; otherwise, the data could be rendered obsolete too rapidly, as technology changes so rapidly. Michael Stebbins, of the White House Office of Science and Technology Policy (OSTP), discussed at length the White House open data initiative, which leverages existing and emerging information channels (such as journal publications) to create data repositories of the future. Another participant suggested that DOD assess the cost of storage vs. reproduction of the data to determine if some sort of data repository (whether mandated or voluntary) makes sense. Yet another participant pointed out that it is fairly common for universities to have permanent storage facilities available, using the Deep Blue program at the University of Michigan as an example. However, other participants pointed out that these programs are expensive for the host and do not always include metadata.

- *Automatic data capture.* Several participants briefly mentioned the use of automatic methods to capture raw and unstructured data from journal articles or to capture and upload raw data from instruments during an experiment. Dave Shepherd, of the Department of Homeland Security (DHS), discussed an automatic data capturing project in biosecurity known as algorithms for analysis, an algorithm-oriented project to identify emerging technologies that could be used against our nation. It uses natural-language processing software to find descriptors in the scientific literature, patents, and other scientific documentation. The data must be up-to-date and continually harvested. Such a project might be useful for the materials community as well.

DATA SIZE: "BIG DATA" VS. DATA

Several speakers noted the recent overall rise in scientific data output. For instance, Mr. Crichton showed that the amount of data produced at JPL has been increasing at a significant and highly nonlinear rate: In 2000, a planetary mission data set contained 4 TB of data. Now, such a project consists of over 500 TB of data. During their presentations, Jed Pitera, of IBM, Mr. Shepherd, and Chuck Ward, of the U.S. Air Force, all noted huge increases in data rates in their communities as well.

However, multiple participants repeatedly questioned whether materials science truly has a "big data" problem or whether it simply has a "data" problem. Several participants noted that the growth in data in materials science, while substantial, is not as dramatic as it is in other scientific disciplines. Ms. Swink argued that important data issues in materials science, such as proprietary data access and the lack of homogeneity, are data problems but not "big data" problems. Other participants indicated that, regardless of the regime, the amount of data produced in materials science is outpacing the algorithms and processing needed to analyze it. They pointed out that the traditional model of data analysis, which relies on the individual researcher, is not likely to be a viable model as the data size continues to increase. At the local level, there may be more data than the researcher has time to analyze. This could indicate need for more centralized analysis and computing tools. Several participants suggested looking at successful big data analysis techniques used in other domains, such as protein databases and work flow models.

Dr. Pitera suggested that the materials science community may be well served by using data reduction or extraction techniques so as to exit the big data regime—in other words, to make the "haystack" smaller and the search for the needle easier. (The idiom "to find a needle in a haystack" was taken up by participants and repeated many times over the course of the workshop.) However, other participants wondered how to perform data selection in the most judicious and domain-specific manner. One person mentioned the need to determine the necessary approach and its associated trade-offs: that is, whether one must interface with a large mass of data or whether one needs algorithms to select the salient features of the data one needs. In a discussion session, Jesus de la Garza, of Virginia Tech, pointed out trade-offs between collecting all the data possible (even if we don't know what to do with them) and collecting only the information we know we want. He observed, however, that most research decisions require an assessment of trade-offs, and this is no different.

QUALITY AND VERACITY OF DATA AND MODELS

Many of the participants referred to the "four V's" of big data (volume, velocity, variety, and veracity)[1] several times during the course of the workshop. Although no one ranked them in importance with respect to materials science, much discussion time was devoted to the last two—variety and veracity. Individual participants noted that there are many sources of uncertainty in materials data, with no consistent methods to verify data quality. This problem underscores the benefit of openly sharing data, so that they can be verified independently.

Dr. Pitera pointed out that data quality can be limited by instrument quality as well as by interpretation quality, and it can be influenced by various data artifacts. Mr. Crichton argued that peer review of data is necessary to assess usability; the international community should agree to standard models and to a consistent peer review process. He suggested that research supported by taxpayer dollars should make its data available and citable: Citations would give the reader confidence that the data set is publication-quality. Several participants discussed challenges associated with the verification of proprietary data. Someone noted that journal publishers are a good point of leverage in the academic community; however, in disciplines such as pharmaceuticals and some others, publication is not a high priority. The data in those instances are used to build business and are considered proprietary. It becomes difficult, if not impossible, for the outside community to gauge the quality of those data.

Several workshop participants complained that, in the materials science community, it is difficult to compare the results of simulations to physical measurements. To help resolve this, Jesse Margiotta, of DARPA, noted that Integrated Computational Materials for Engineering (ICME) has a vigorous verification and validation component used to provide confidence limits. Another participant suggested using a materials work flow platform (such as Kepler[2]) to capture data and enable reproducible results.

---

[1] For a full description of the four V's, see the section "IBM and Big Data" in Session 1.
[2] Kepler is a free, open source, scientific work flow tool. See http://kepler-project.org for more information. Accessed February 25, 2014.

DATA AND METADATA ONTOLOGY AND FORMATS

Another challenge lamented at the workshop was the lack of standards and terminologies for data and metadata. Several participants noted the absence of a formal ontology for materials science and the need for a practical set of identifiers and descriptors. It was noted during a discussion that the materials community may suffer from a dearth of conversation about ontology.

One participant complained that the process for developing standard terminology is very difficult and slow, pointing to the erstwhile ASTM International committee that once developed standards in this area. In general, however, companies do not like to pay their employees to do this type of activity, and the ASTM International committee folded because there was no funding for it from the community. The same participant noted that companies may not be interested in attaching themselves to a certain standard format, because they are concerned they will be forced to share information they would prefer to keep proprietary in their own formats. During his presentation, Mr. Crichton suggested that the international community should agree to standard models and to a consistent peer review process. He pointed out that agreement on how to represent data can be very difficult because different scientists will have different emphases within the same data set. Mr. Shepherd also pointed out problems with many formats, resolutions, and source locations of large data sets.

To move forward, one participant suggested looking to the NSF program EarthCube[3] as a model for working across different communities to develop ontologies and names. Dr. Margiotta also reported that DARPA, along with the Army Research Laboratory (ARL) and other program partners, is developing methods to standardize data fields and metadata fields for materials and materials processing.

METADATA AND MODEL AVAILABILITY

The concept of metadata availability had several meanings at the workshop. Mainly it referred to access to knowledge and information about a particular experiment—models used, starting conditions, and other "meta" information about the data—in other words, information that is not generally available in a journal article, which limits one's ability to replicate the experiment. Many participants spoke over and over about the need to capture and report metadata.

In other cases, metadata availability referred to broader access to the models themselves; there were several separate discussions about the need to have a stan-

---

[3] The NSF EarthCube program is an integrated data management program in the geosciences. See https://www.nsf.gov/geo/earthcube/ for more information. Accessed February 24, 2014.

dard modeling toolkit available to researchers. A participant suggested that, for the small-scale researcher, data production is outpacing computing capabilities. In the case of neutron beams, for example, the bottleneck is in processing data, not in having access to beam lines. Continued progress is needed in developing new data analysis programs. Several participants and speakers discussed the idea of transitioning data computation to the source of the data. (This is also related to the workshop discussions of data size, summarized above.) Mr. Crichton said that this paradigm is likely to become increasingly important in the next 5 years, as data sets become more and more massive. More analysis tools and capabilities will be needed at the site of the data repositories. Mr. Crichton predicted that merely distributing data would soon become an obsolete approach; instead, research services and analytics will be a more advantageous approach, as users will need services rather than just the data.

Mr. Shepherd, of DHS, described a project in predictive biology that uses knowledge-based (KBase) data to combine data for microbes, microbial communities, and plants into a single integrated data model. Users can upload their own data and build predictive models, which represents more of a community effort in big data. Its plug-in architecture will allow other laboratories to use their own algorithms to analyze the data. This project mixes big data, high-performance computing, and cloud architecture. This type of community modeling effort may be a useful example for the materials science community.

Several participants stressed the importance of developing and distributing models for predictive purposes, particularly to determine inspection and maintenance intervals and life prediction as a function of use. Dr. de la Garza stated the importance of prediction rather than reaction; he suggested that big data analytics for prediction would be a powerful tool in the materials community. Mr. Shepherd and Dr. Pitera both provided examples of predictive modeling in different areas (biosecurity and equipment maintenance, respectively).

CULTURE

Several participants noted that the materials community lacks a data-sharing culture. This is likely the result of a variety of factors, they believed. The most prominent factor is the reluctance, particularly on the part of private industry, to share proprietary data or other data that may have value. One participant noted that withholding materials data is practiced in order to gain a competitive edge, and companies want to protect their intellectual property. Another factor discussed was the difficulty associated with working under sometimes complex DOD contract rules and regulations.

Several participants discussed the need for incentives for sharing data. They proposed specific incentive models for sharing and structuring materials data in lieu of purchasing data ownership rights. Such incentives included the use of data citations, the development of quid pro quo relationships, and the creation of precompetitive partnerships. Some participants suggested using existing information channels to increase the opportunity for data sharing, such as linkages from journals to data repositories, consistent with the goals of the White House initiative on openness and access to scientific data.

Session 1: Introduction to Big Data

Session 1 of the workshop focused on introducing the concepts of big data and their application to several different disciplines. Presentations were made by Dan Crichton, Jet Propulsion Laboratory (JPL), Jed Pitera, IBM Research-Almaden, and Dave Shepherd, Department of Homeland Security (DHS).

FRONTIERS IN MASSIVE DATA ANALYSIS AND THEIR IMPLEMENTATION

Daniel Crichton, Director, Center for Data Science and Technology, Jet Propulsion Laboratory

Mr. Crichton explained that he manages multiple projects and programs in scientific data systems for planetary, Earth, and other disciplines. JPL has a data science and technology center that supplies the infrastructure needs and end-to-end systems for scientific data systems across multiple disciplines. Mr. Crichton said that he was also a member of the committee that wrote the NRC report *Frontiers in Massive Data Analysis* (2013) and would describe some of the findings and recommendations of that study.

Mr. Crichton stated at the outset that all observational science platforms, whether from a Mars rover, an Earth observing system, or a neonatal sensor system for an infant, have certain elements in common: All simultaneously provide numerous forms of measurement and observation. He noted that there are several challenges in space systems, and probably in other domains as well:

- Space systems are developed and deployed worldwide, and data are generated across complex, interconnected systems. Mr. Crichton said that there are many producers of the data; as a result, the data are highly distributed, with limited data sharing.
- Systems are heterogeneous and located in different physical systems. As a result, the data are stored in a variety of formats, and access to data can be difficult.
- The data sets are so massive that they stress traditional analysis approaches.

Mr. Crichton explained that the elements to collect data are traditionally built independently, with each element supporting a single type of data and data collection. The concept of big data is the opposite: Big data seek to look at the end-to-end architecture needed to support an entire system. The big data approach considers ways to scale and systematize analyses. He showed that the amount of data produced at JPL has been increasing at a significant and highly nonlinear rate.

Mr. Crichton explained the importance of an architectural approach to big data. He said that the National Aeronautics and Space Administration (NASA), for instance, had captured its data in high-quality archives, but NASA did not originally consider the full data life cycle; instead, it pushed the burden of data analysis to the individual researchers. In contrast, the big data approach supplies a technical infrastructure with an overall architecture. In addition, Mr. Crichton stressed that the technical infrastructure should be built independently of the data and their structure. In general, technology changes rapidly, and data should not become obsolete on the same rapid time scale as the technology. Once the infrastructure is in place to support data capture, it is also important to be able to analyze the data. Mr. Crichton suggested the adoption of more advanced techniques for data analysis to increase the efficiency of data analysis in a distributed environment. He gave the example of a space mission with an interferometer. In the past, such a mission would provide a few hundred terabytes of data; now that same mission format will provide tens to hundreds of petabytes of data. This data explosion requires changes to data storage, processing, and management.

Mr. Crichton briefly described the big data life cycle, from data generation to archiving and analysis. In between, data triage is conducted; if the data are too massive, data reduction steps may be necessary to reduce their overall size. However, any data reduction requires inferences to be developed about the data, adding uncertainty to the data.

Mr. Crichton pointed out that NASA has specific challenges to the standard data collection paradigm. He defined three challenges that are encountered in space data systems in particular:

1. How do we store, capture, and transmit data from extreme environments?
2. How do we triage massive data for archiving?
3. How can we use advanced data science methods to systematically derive scientific inferences from massive, distributed science measurements and models?

Mr. Crichton said that NASA's data are collected by onboard instruments, and NASA is limited as to how much data can be returned to the ground. It is therefore interested in onboard computing to perform data reduction. However, the science community is not always in favor of data reduction. In any case, onboard data collection systems are likely to reach a capacity constraint in the near future, which will force a change in system implementation.

Mr. Crichton then listed a set of technology trends identified by the NRC report on massive data (NRC, 2013):

- *Distributed systems.* This trend includes different access mechanisms and ownership rules, data federation, linking, etc.
- *Scalable infrastructures and technologies to optimize computing and data-intensive applications.* Mr. Crichton noted that the trend is moving toward data-intensive computing rather than high-performance computing.
- *Service-oriented architectures.*
- *Ontologies, models and information representation.* Mr. Crichton stated a need to agree on how to represent data; this can be very difficult, as different scientists will have different emphases within the same data set.
- *Scalable database systems with different underlying models.*
- *Federated data security mechanisms.* Different access permission rules will create both authentication and authorization issues.
- *Technologies for the movement of large data sets.* Mr. Crichton pointed out that this is a big challenge in the high-energy physics community in particular.

Mr. Crichton then described the framework for data-intensive systems built at NASA's JPL. The framework is open source, and it has been applied in various settings by the broader community. He said that there is no single big data solution; instead, one must bring the different building blocks together for each problem in such a way that they can be scaled and optimized. The framework is known as Object Oriented Data Technology (OODT).[1] OODT was originally developed to support NASA's Planetary Data System (PDS), a system recommended by an earlier NRC study. The data were stored on tapes, and the quality was beginning to erode; an NRC study recommended the development of the PDS (NRC, 1995).

---

[1] OODT can be explored online at http://oodt.apache.org/. Accessed February 20, 2014.

The purpose of the system is to collect, archive, and make accessible digital data and documentation produced from NASA's exploration of the solar system. Mr. Crichton explained that the data have international operability and have been validated against a common set of structures and data standards.

Mr. Crichton then described the earth science data pipeline: Systems are built to capture data. Those data are sent to the ground station; higher-order data products are developed; the results are archived; and the results are provided to the community at large. This system has worked well; currently, it consists of a well-curated repository of about 7 PB, though that number will increase rapidly. Mr. Crichton said that JPL is also supporting the Earth System Grid Federation, which informs the entire earth science modeling community and also supports the Intergovernmental Panel on Climate Change (IPCC) assessments.

Mr. Crichton then said it is important to evolve from data archiving to scalable data analytics. Data analytics requires automated methods to identify and detect data patterns across a variety of disciplines and under many different operational paradigms. He pointed out that a system could be used to study Twitter patterns for security purposes, for example. Mr. Crichton said that JPL is focusing on the transition from the data computation to the source of the data. This paradigm is likely to become increasingly important in the next 5 years, as data sets become more and more massive. More analysis tools and capabilities will be needed at the site of the data repositories. Mr. Crichton predicted that merely distributing data would soon become an obsolete approach. Instead, research services and analytics will be a more advantageous approach, as users will need services rather than just the data.

Mr. Crichton explained the elements to enable big data analytics for NASA. A schematic of these elements is shown in Figure 1. Currently, there are a variety of data sources, including satellite archives, airborne instruments, in situ data, and ground sources. These constitute a massive data set that has been brought together but that was not originally designed to be integrated. The data science infrastructure, such as the data, algorithms, and machines, informs data analytics. Big data triage includes the following elements:

- *Detection.* Identify elements of interest.
- *Classification.* Organize data automatically and in real time.
- *Prioritization.* Use the information to inform adaptive data compression.
- *Understanding.* Explain events for humans to understand and interpret the results.

Mr. Crichton pointed out that the same challenges exist in other fields, not just planetary science or earth science. For example, cancer research databases will probably require similar automatic processing of data, and astronomy and planetary science will also likely need automatic processing and feature detection.

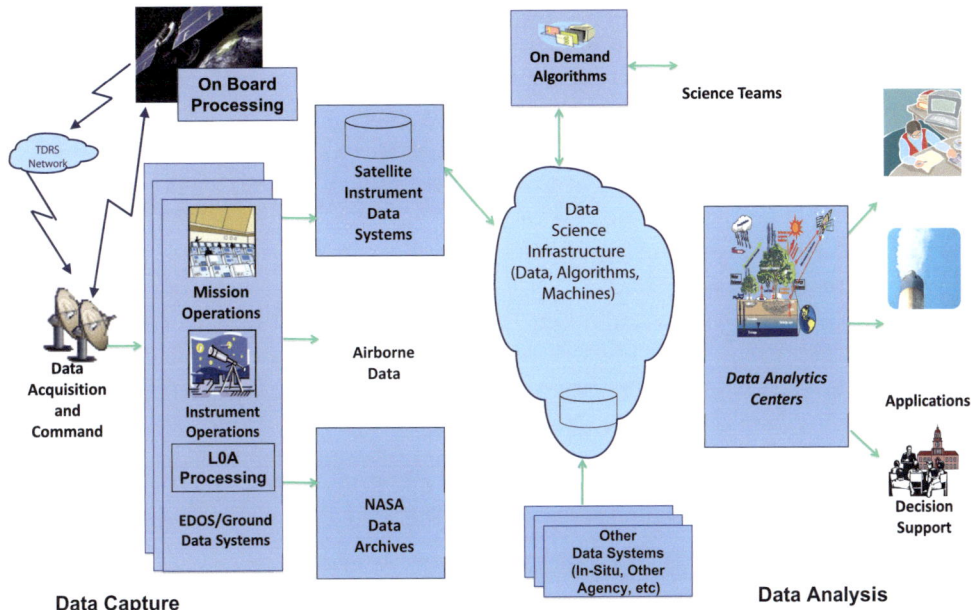

FIGURE 1 The elements of systematic analysis of massive data for NASA. SOURCE: Dan Crichton, Jet Propulsion Laboratory, presentation to the committee on February 5, 2014, Slide 17.

Mr. Crichton then explained that the 2013 NRC *Frontiers* report emphasized how to systematize and automatically analyze data. The report also stressed the need to bring together experts in multiple disciplines to solve these big data problems.

Mr. Crichton concluded with several of his own recommendations:

- Invest in capturing and maintaining data in well-annotated, accessible, structured data repositories.
- Bring together cross-disciplinary groups to understand how to systematize massive data analysis and increase efficiencies.
- Develop computing architectures for sharing and analyzing highly distributed, heterogeneous data, aided by international cooperation and coordination.
- Enhance sustainability in both the data and the technology infrastructures while keeping the two on separate paths to maintain the long-term health of the system.

In the discussion period, several participants suggested mapping these ideas onto advanced materials and manufacturing. Some other participants noted that the materials and manufacturing communities may not even have the models to analyze the data; in this case, care has to be taken not to "throw the baby out with the bath water" during the data triage process. In the materials community, it is currently very important to capture the raw data. A few participants discussed the need to (1) understand the best way to downselect the data rather than to capture all of it and (2) focus on models to best understand the salient features of the data that are necessary to retain.

One participant asked about data repositories and how to ensure that data remain accessible in the future. Mr. Crichton responded that it is important to separate the data from the technology. Data need to be captured in formats that are stable, even if that means that the format is not contemporary. The data need to be software independent and to rely upon conceptual models that are static. Mr. Crichton said that archives typically require at least a 50-year expected time window.

Mr. Crichton was also asked about how to address differing data quality. He responded that peer review of data is necessary to assess usability, and that the international community needs to agree to standard models and to a consistent peer review process. He said that federal research should make its data available and citable: Citations allow the reader to have confidence that the data set is publication-quality.

IBM AND BIG DATA

Jed Pitera, Manager, Computational Chemistry and Materials Science, IBM Research-Almaden

Dr. Pitera explained that he manages IBM's research team in computational chemistry and materials science at IBM Research-Almaden. He said that this group generates large volumes of structured data and that it focuses on computation more than on data. His research group examines computational problems in soft condensed matter (i.e., polymeric materials), which includes the computational screening of chemical reactions, the development of multiresolution algorithms, models for future technologies for lithography and nanopatterning, and models of fundamental interactions in chemistry and physics. In addition to his group's data needs, Dr. Pitera pointed out that IBM also must analyze large amounts of unstructured data to support its business applications.

To add nuance to the word "big" in big data, Dr. Pitera described some of its specific challenges. He stated that big data are often described in terms of the four V's:

1. *Volume.* Volume refers to the physical amount of data. Dr. Pitera stated that problems related to volume in big data are scaling and the cost of dealing with that much information as much as structural problems.
2. *Velocity.* Velocity is both the rate at which data are collected and the time between receiving input and requiring an output. Dr. Pitera gave several examples of different time frames: In some instances, the maximum duration is several minutes (such as the time scale of a "coffee break," the time it takes for a researcher to get a coffee and return to work); in other instances, the maximum duration is a few seconds (such as the time a call center operator might need to return an answer to a client).
3. *Variety.* Data come in a wide variety of formats.
4. *Veracity.* There are many sources of uncertainty associated with data quality. Data can be limited by instrument quality, as well as interpretation quality, and data can be influenced by various data artifacts.

Dr. Pitera stated that managing the four V's requires expertise in data (acquisition, integration, cleansing, storage, protection, and management), domain (models and hypotheses), informatics (algorithms, simulations, and rules), mathematics (analysis), systems (scale and velocity), and visualizations (sharing results). He said that his group's efforts focus primarily on the first two areas of expertise (data and domain): How do we get data, ingest them, and structure the domain of interest to capture the needed information? He posited that there are opportunities for big data in all types of study. The discovery phase focuses on conducting computational experiments, mining the literature (the published literature, as well as unpublished laboratory documentation), and finding new materials or repurposing existing ones. Dr. Pitera postulated that people tend to focus on the discovery phase. However, the material must be made usable, and integration remains a challenge. Life-cycle issues, including continuous reengineering and monitoring, are all topics to be explored.

Dr. Pitera described three representative projects in materials and big data:

- *Harvard's Clean Energy Project.* This is a distributed computing project that focuses on using big data for materials discovery. It seeks to identify new photovoltaic materials with higher efficiency rates. This involves large-scale calculations, along with data mining and analytics.[2]
- *Pharmaceuticals.* Pharmaceutical companies wish to mine patent data and medical literature to identify new relationships between drugs and disease.
- *Mining equipment.* The goal of this project is to understand predictive equipment maintenance for heavy mining machinery.

---

[2] See http://cleanenergy.molecularspace.org/about-cep/. Accessed June 2, 2014.

Dr. Pitera noted that the latter two projects are being developed in IBM's Accelerated Discovery Laboratory, an exploratory collaboration space for experimental big data projects that include government and academic and other private partners.

In closing, Dr. Pitera reminded participants that materials are complex, more than just a set of chemical reactions. Understanding a material requires knowledge about its statistical composition, processing history, and other metadata. He said that systems to manage data improve daily and asked if an IBM Watson[3] for materials would be useful. He pointed out that if the same questions are asked repeatedly, then a Watson would be useful and the development costs could be justified. If not, what other type of infrastructure could be used? Finally, Dr. Pitera suggested that big data should perhaps be thought of as useful data that just happen to be big. He said that it may make sense to implement data reduction or extraction to leave the big data regime—in other words, it may be better to make the haystack smaller.

During the discussion period, a participant brought up a challenge unique to materials science: A material is qualified based on its characteristics, not its composition. Two materials may qualify for the same application but have different compositions. Examples such as aluminum alloys and jet fuel were discussed. Participants also wondered what the community should do next to come together. Dr. Pitera suggested that effort should be put into the construction of appropriate data models. These models are likely to be domain-specific, as metallurgy differs from polymer chemistry, which in turn differs from inorganic glass. A uniform description for data models would need to involve many different communities.

BIG DATA FOR BIOSECURITY

Dave Shepherd, Program Manager, Homeland Security Advanced Research Projects Agency, Department of Homeland Security

Mr. Shepherd began by noting that he works in biology programs with homeland security applications and that his portfolio does not include materials or manufacturing. He explained that the goal of biosecurity is to alleviate any accidental or intentional release of pathogens or other causes of disease. He said that biosecurity is a predictive field, seeking to prevent disease exposure and spread by means of anticipation, and it requires a long-term perspective. He said that this predictive need was identified by an OSTP workshop in 2013.[4] At the OSTP

---

[3] Watson is an artificially intelligent computer system capable of answering questions posed in natural language. At one point Watson had access to 200 million pages of structured and unstructured content in a database consuming 4 TB of disk storage. See http://www.pcworld.com/article/219893/ibm_watson_vanquishes_human_jeopardy_foes.html. Accessed June 2, 2014.

[4] www.whitehouse.gov/sites/default/files/microsites/ostp/biosurveillance_roadmap_2013.pdf.

meeting, three of the five panel topics were related to big data (Big Data for Early Indications and Warnings, Big Data for Digital Surveillance/Digital Disease Detection, and Big Data Research and Analytics), and many of the problem sets discussed included the need for predictive tools.

Mr. Shepherd next explained that big data permit computer-assisted analytics, and algorithms and analytics can be used to classify and identify important information on a massive scale. Big data allow for correlations on a level that would not be possible with a smaller subset of information. One additional advantage for biosecurity is that surrogate data can be used. However, biosecurity big data sets tend not to be clean and do not contain consistent metadata. In addition, it can be difficult to find the right biosecurity data set for one's needs.

Mr. Shepherd then went on to discuss three specific use cases for big data in biosecurity:

- *Use Case 1: Algorithms for Analysis.* This is an algorithm-oriented project to identify emerging technologies that can be used against the United States. It uses natural-language processing software to find descriptors in the scientific literature, patents, or other scientific documentation sources. The data must be up-to-date and continually harvested. Permissions are needed to access all the different possible data sources.
- *Use Case 2: Size and Architecture.* This project uses knowledge-based (KBase) data for modeling for predictive biology. KBase data combine data for microbes, microbial communities, and plants into a single integrated data model. These data are considered in a single, large-scale bioinformatics system. Users can upload their own data and build predictive models. This endeavor represents more of a community effort in big data. This project currently consists of 1 PB of stored data, and that amount is likely to increase significantly. Its plug-in architecture will allow other laboratories to use their own algorithms to analyze the data. This project mixes big data, high-performance computing, and cloud architecture.
- *Use Case 3: Surrogate Data.* Because the actual emergence of a novel biosecurity threat is rare, researchers seek to find surrogate data to help when new threats are encountered. Mr. Shepherd believed that the use of surrogate data is very helpful to this community, but that the idea of surrogate data may not be feasible in other communities. One example where excellent surrogate data for biosecurity exists, Mr. Shepherd noted, are data for the spread of antimicrobial resistance. Because outbreaks of antimicrobial resistance occur rather frequently, there are enough data to support predictive model development and verification. Like antimicrobial resistance, a novel biological threat can appear anywhere in the world, so

infrastructure to support the development of a geographic distribution map would be useful and would assist in analysis and prediction.

Mr. Shepherd pointed out that clinical data are not aggregated at a national level, and there is no infrastructure to support such data aggregation. He said that researchers are beginning to realize that a centralized database is unlikely to be viable. When human subjects and privacy are involved, data sharing becomes difficult. Instead, it may make the most sense to move the data processing to the individual holders of clinical data. Mr. Shepherd envisions a new model in which clinical data holders participate in a national, distributed, interconnected grid similar to the collaborative model that underlies Lawrence Livermore National Laboratory's Earth Systems Grid Federation for the climate modeling community.

Mr. Shepherd then described some of the challenges associated with big data in biosecurity, noting that these probably apply to other fields as well:

- *Data quality and access.* There are many formats, resolutions, and source locations. The use of images increases the data storage and transmission needs.
- *Personal data.* There are ongoing concerns related to personal data, privacy, and civil liberties.
- *Analytics.* Algorithms are not currently keeping pace with the enormous amount of data. More analytic processing capability is needed.

Mr. Shepherd concluded by saying that increased control over information has not led to increased biosecurity; instead, in the future, increased information sharing may be the way to increase biosecurity.

In the discussion, a participant suggested that it might be better to store the actual material instead of data describing the material. This brings up practical issues associated with how to store materials. This is true in biosecurity as well: Should you store the biological agent, data about the agent, or both?

A participant asked if the human immunodeficiency virus (HIV) and severe acute respiratory syndrome (SARS) would make good surrogates in biosecurity. Yes, said Mr. Shepherd, because emerging diseases, even ones that don't represent a typical biosecurity scenario, can still provide modeling information.

Dr. McGrath pointed out that biosecurity is a large and difficult problem, larger in scope than what the materials community faces. He suggested interfacing with other communities that use predictive tools, such as the intelligence community. On the science side, he suggested working on sensor development and increasing basic knowledge and understanding.

A participant pointed out that cloud computing can be considered high-performance computing as well. Mr. Shepherd agreed and explained that he divided them in his discussion so that high-performance computing referred to anything from chips to run times, and that cloud computing referred to how to work together.

DISCUSSION

Ms. Swink opened the discussion session with a summary of several statements she had found compelling. One was the importance of determining the necessary approach and the associated trade-offs: whether one needs to interface with a large mass of data, or whether one needs algorithms to select the salient features of the data one needs. She also brought up the concept of ontology and asked whether this is a driver in the materials research and development area. Finally, she asked if the idea of surrogate data had any analog in the materials and manufacturing community.

Dr. de la Garza pointed out the need for a holistic perspective on big data, because they involve not only data collection but also life-cycle information: data collection, procurement, analytics, and stable, archival-quality formats. He also noted that there are trade-offs between collecting all the data possible (even if we don't know what to do with it) and collecting only the information we know we want. He observed, however, that most research decisions require an assessment of trade-offs, and this is no different. Dr. de la Garza also underlined the importance of prediction over reaction; big data analytics for prediction would be a powerful tool.

Some participants also discussed the idea of storing materials vs. storing data. Ms. Swink asked the group to consider the reestablishment of critical material repositories within DOD. She indicated that she is aware of a reluctance to do so. She also argued that there is a large financial difference between having stockpiles or archives: The workshop participants were more likely to be interested in the less-expensive archives. One participant mentioned the Air Force's digital twin program, in which the digital representation of a material keeps information about the material properties; perhaps it would be valuable to include an actual sample to examine along with the digital twin. However, other workshop participants also pointed out that critical policy issues would need to be addressed, such as the amount of material to retain, access criteria, and other issues. Others argued that any sort of repository, even an archive, might be prohibitively expensive. It may be necessary to set priorities on which materials to retain.

Participants also discussed common elements across disciplines that rely on big data. Some of these common challenges are associated with moving data, connecting to different ontology models to describe the data, and tools to identify compounds. Another popular idea across disciplines was bringing the processing to the data source and other distributed data strategies. A participant noted that a Hadoop[5] might play a useful role in distributed data storage and processing as part of a larger solution.

A participant suggested that the materials science community look at examples in other disciplines to identify the costs associated with data storage. The participant suggested the fields of publishing, which is moving toward open access, and evolutionary biology, which has a significant amount of data but no repository structure. Another participant noted the utility of the open source model used by Dryad to provide access to data.[6]

Ms. Swink asked what steps the community should take now to move forward in data management. One participant suggested looking to the NSF program EarthCube as a model for how to work across different communities to develop ontologies and names.[7] The materials community may suffer from the lack of conversation about ontologies.

Ms. Swink then asked if the materials science community has a "data" problem or a "big data" problem. She suggested that the problems are just data problems. She argued that important data issues in materials science, such as access to proprietary data and the lack of homogeneity, are not big data problems. She also pointed out that the long-time separation between materials development and product commercialization interferes with a company's ability to understand the correlation and causality between product success and materials data.

---

[5] Hadoop, developed by the Apache Software Foundation, is an open-source software framework for the storage and processing of large-scale data sets on clusters of distributed computers. See http://hadoop.apache.org/ for more information. Accessed February 20, 2014.

[6] The Dryad Digital Repository is an open source to make available the data that underlie scientific publications. See http://datadryad.org for more information. Accessed February 24, 2014.

[7] The NSF EarthCube program is an integrated data management program in the geosciences. See https://www.nsf.gov/geo/earthcube/ for more information. Accessed February 24, 2014.

Session 2: Big Data Issues in Materials Research and Development

Session 2 of the workshop focused more specifically on the application of big data to materials research and development. Jim Davis, University of California Los Angeles, had originally been scheduled to present during Session 3, but the workshop participants determined his talk content was well suited to the discussions in Session 2, so he was moved here. Thom Mason, Oak Ridge National Laboratory; Chuck Ward, U.S. Air Force; Rusty Irving, GE Global Research; and Dr. Davis made presentations in Session 2.

PHYSICS IN BIG DATA

Thom Mason, Director, Oak Ridge National Laboratory

Dr. Mason began by noting that the title of his talk might be misleading. His presentation focuses on scientific data, he said, which includes physics, but it includes many other fields as well. He explained that Oak Ridge National Laboratory (ORNL) is a DOE laboratory, and its drivers are derived from the DOE mission. Many of these drivers involve large amounts of data and computing. He pointed out that the data requirements vary across scientific disciplines. High-energy physics and cosmology have very high data volumes, on the order of 1 PB per day. Data volumes for neutron sources are somewhat smaller, but they are still bigger than can be comfortably managed. Improvements in imaging systems have resulted in larger volumes of data, the scale of which can overwhelm a smaller research group with access to an ORNL beam line. Dr. Mason said that data come in many varieties (such imaging data, text, and sensor data), and the content must be combined. He argued that this growth in data is outpacing computing growth.

Dr. Mason also noted, as an aside, that a particular challenge in the materials community is that of provenance. Other disciplines, such as cosmology, have no concept of a sample. In materials, individual researchers invest much time and effort in producing and characterizing a sample, and as a consequence there will be challenges associated with moving to open source data because of the intellectual investment in a particular sample. The variability of samples between the different groups producing them is another challenge.

Dr. Mason stated that ORNL is DOE's largest science and energy laboratory, and materials science permeates all of ORNL's focus areas. He described ORNL's high-performance computing ecosystem, including its flagship system, known as Titan.[1] ORNL's high-performance computing system models and simulates physical, biological, and chemical systems. It processes data, but it generates large volumes of data as well. The primary drivers for the supercomputing power at ORNL are scientific, but ORNL is also exploring industrial applications. Dr. Mason provided three examples:

- *General Motors.* General Motors sought to understand and develop new thermoelectric materials for higher fuel efficiency.
- *BMI.* BMI Corporation[2] sought to create parts to retrofit onto Class 8 long-haul trucks for improved fuel efficiency and emissions.
- *Boeing.* Boeing sought to develop and validate computational models for transport airplane design and development.

Dr. Mason then described the Accelerating Data Acquisition, Reduction, and Analysis (ADARA) program at ORNL. ADARA is an experimental data analysis program developed in response to increased data and imaging needs from the Spallation Neutron Source. He pointed out that the Spallation Neutron Source is improving, and data volumes are growing at corresponding rates. In the past, the data volumes were handled via an individual researcher's analysis code. The general rule of thumb was that one week of data collection equaled one year of graduate student analysis time. Now, however, the data are over a hundred times larger in size, and the old model of analysis has become unsustainable. When Dr. Mason was later asked if ADARA was limited to the Spallation Neutron Source, he clarified by saying that while ADARA is specific to the Spallation Source, the concept

---

[1] More information about Titan can be found at http://www.olcf.ornl.gov/titan/. Accessed March 10, 2014.

[2] BMI Corporation is an engineering services firm based in Greenville, South Carolina.

is applicable to other neutron and X-ray sources. The code is largely open source, facilitating expansion elsewhere.

Dr. Mason also described ORNL's Manufacturing Demonstration Facility (MDF), which includes extensive capabilities for additive manufacturing. ORNL is interested in expanding the palate of materials it uses for additive manufacturing. The laboratory is developing titanium alloys (primarily for the medical device community), high-temperature alloys, and refractory metals (such as tungsten). The processing recipes and real-world performance of these materials are still unknown. Dr. Mason pointed out that ORNL's high-performance computing capabilities are useful to model the additive manufacturing of complex shapes. He then described several examples from the MDF:

- *Additive manufacturing of turbine blades.* ORNL and its industrial partner, Morris Technologies, use neutron scattering to measure atomic plane spacing and better understand the link between residual stress and laser additive manufacturing processing. They seek to develop a pedigree that will enable the turbine blades' use even in situations that could otherwise compromise human or environmental safety.
- *Rapid, agile manufacturing.* ORNL and its industrial partner, Local Motors, use crowdsourcing to design vehicles for short-run manufacturing. The project focuses on novel material development and additive manufacturing. They are currently creating a process for large-scale (20 feet per side) carbon-reinforced polymer development that combines additive and subtractive techniques.

In the discussion period, Robert Schafrik, GE Aviation, asked about the additive manufacturing processes. He pointed out that the construction of a large structure takes time and will require changes to the additive process. Those changes might affect the properties of the end result and should be accounted for in the models. Dr. Mason pointed out that ORNL is moving toward parallel processing with large component sizes. One promising avenue to improve throughput is to use additive manufacturing for tool and die production, greatly speeding up the development cycle, while employing traditional high-throughput techniques for volume manufacturing.

The discussion briefly turned to access to data. Dr. Mason said that ORNL desires to move to an open source model. Another participant thought that embargo periods might be useful; access to data could be restricted for a period of time, then the data could be sent to a public repository. Another participant discussed the problems associated with a protein crystal structure databank; researchers are required to submit data to that databank. The community is discovering that much of the data is wrong, a sign that experimental constraints may need to be imposed for data consistency. Dr. Mason explained that, for the vast majority of work done at ORNL, the user owns the data but the expectation is that the data will be published. Under those constraints, ORNL does not collect a user fee. A small amount of work conducted at ORNL uses proprietary data, and that work is arranged under a full-cost recovery model.

MATERIALS GENOME INITIATIVE AND BIG DATA

Charles Ward, Lead for Integrated Computational Materials Science and Engineering, Air Force Research Laboratory

Dr. Ward began his presentation by showing that the amount of information related to materials has been increasing dramatically each year; in 2012, over 162,000 journal articles were published in the fields of materials science and engineering. He pointed out that industry has some of the best materials databases in the world. He gave an example of data-driven materials development in a cast and wrought disk alloy, R65, conducted by GE Aviation. Data-driven methods such as these are able to reduce both development time and costs by up to 50 percent.[3]

Dr. Ward then described the Materials Genome Initiative (MGI), which intends to create a Materials Innovation Infrastructure comprising accessible digital data, computational tools, and experimental tools. More about the MGI can be found in Box 1.

Dr. Ward then defined data and metadata. Data are the result of measurement; data can be physical, consisting of experimental results and uncertainty, or virtual, consisting of simulation results and uncertainty. Metadata are the information that describes the measurement process; metadata can also be physical, consisting of the experimental setup and settings, or virtual, consisting of the explicit underlying model, simulation software, and input parameters.

Dr. Ward presented an example of a traditional approach to materials development that focused on measuring a diffusion coefficient. In the example, a researcher would measure the diffusion coefficient by tracking the root mean displacement of tracer particles, record the values, and publish the results (including metadata in the publication), typically represented as a diffusion coefficient. However, this reporting approach is dependent on a specific model (the diffusion equation). The paper assumes that the data fit the model, and it does not actually report the experimentally measured data.

Dr. Ward noted that the crystallography community has moved toward open data sharing; crystallography is one of several disciplines that are leading the way

---

[3] R. Schafrik, GE Aviation, paper presented at the RTO Applied Vehicle Technology Panel (AVT) Specialists' Meeting, *RTO-MP-AVT-187*, on October 12, 2012.

BOX 1
Materials Genome Initiative

The Materials Genome Initiative (MGI), announced by the White House in June 2011, aims to double the speed at which materials are discovered, developed, and manufactured. MGI seeks to develop a materials innovation infrastructure that includes the following:

- Computational tools for modeling, simulation, design, and exploration.
- Experimental tools for synthesis and processing, characterization and analysis, testing and prototyping, and verification and validation.
- Digital data.

The goal of the MGI is to develop collaborative networks that support the sharing of best practices and data to foster an open environment for materials design and development. Areas of impact include clean energy, human welfare, national security, and the next generation workforce (see Figure B-1).

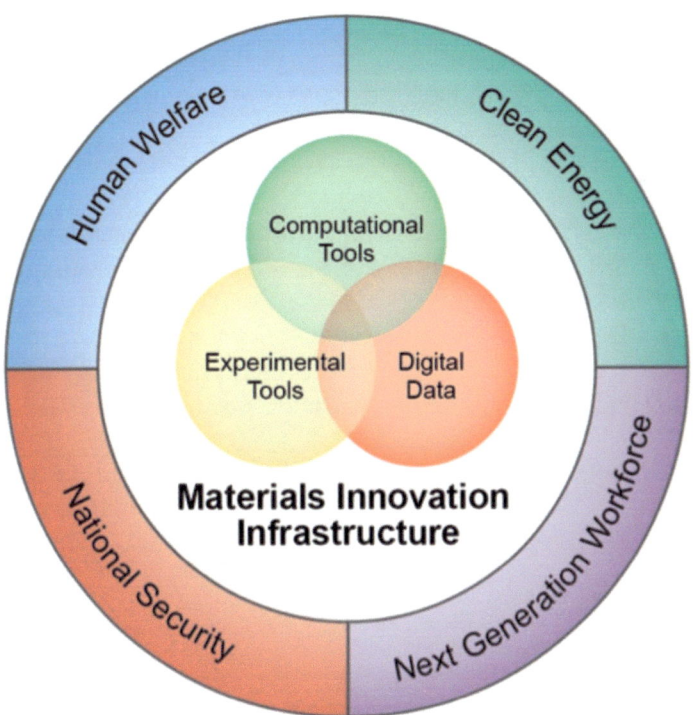

FIGURE B-1 Conceptual representation of the MGI, showing the overlapping infrastructure requirements and the application areas.

SOURCE: White House Materials Genome Initiative, http://www.whitehouse.gov/mgi/goals. Accessed February 26, 2014.

in open source data. The Crystallographic Open Database stores over 250,000 compounds and minerals. The International Union of Crystallography has mandated data archiving for its journals; as a result, one database (Crystmet) has over 150,000 entries of metallic and intermetallic structures, and another (Cambridge Structural Database, or CSD) has nearly 700,000 organic and metal-organic structures and almost 100,000 macromolecular structures available in its database. Dr. Ward suggested a next logical step for the materials community would be to capture data pertaining to CALPHAD.[4] CALPHAD data also consist of metadata that increase in spatial and temporal complexity compared to crystallographic data.

Dr. Ward then posed a set of challenges in addressing materials data. He noted the following:

- *Materials have a pervasive application.* As a result, there is no single government funding agency that leads a cohesive materials research and development effort.
- *Materials have a nearly infinite scale of design variability.* There are few reference data sets, making it difficult to standardize data descriptions.
- *Materials are studied and used by many technical disciplines.* As a result, the data are widely dispersed, and a disparate vocabulary is used to describe them.
- *Materials are often a product differentiator.* This means that proprietary protections are often put into place for materials, making data sharing more difficult.
- *Materials are a key to economic and national security.* As a result, they are subject to export controls, such as the International Traffic in Arms Regulation (ITAR) and Export Administration Regulations (EAR).[5]

Dr. Ward provided information on the National Institute for Standards and Technology (NIST) workshop on MGI data, held in May 2012. The workshop identified common themes to be addressed for materials data archiving, including the following:

- Materials schema/ontology,
- Data and metadata standards,
- Data repositories/archives,

---

[4] CALPHAD refers to the CALculation of PHAse Diagrams, a computational method of modeling the thermodynamic properties of materials.

[5] For further reading, see, for example, NRC, *Export Control Challenges Associated with Homeland Security*, Washington, D.C.: The National Academies Press, 2012, and NRC, *Beyond "Fortress America": National Security Controls on Science and Technology in a Globalized World*, Washington, D.C.: The National Academies Press, 2009.

- Data quality,
- Incentives for data sharing,
- Intellectual property, and
- Tools for finding data.

Dr. Ward also showed results from an open survey conducted by the Materials Research Society and The Minerals, Metals and Materials Society (TMS) in the summer of 2013. The survey had a large number (675) of respondents. The respondents reported an interest in sharing fairly modest amounts of data overall (about half of the respondents said the data quantity they wish to share would be <1 GB per year). Survey participants also responded that they would be most interested in databases and data mining tools related to a material's physical and thermal properties. This is actually the most basic level—physical and thermal properties have the least complex data and metadata requirements. Survey respondents noted some impediments to data sharing, particularly with respect to data ownership and intellectual property rules. One slight surprise was that receiving feedback from others was considered a strong motivating factor for sharing data, on a par with increased research visibility.

Dr. Ward then described MGI initiatives with data-intensive elements taking place within the government, including a number of projects at ARL, the Air Force Research Laboratory (AFRL), ONR, DARPA, DOE, and NIST. He pointed out that these efforts use an ICME framework. (For more information about ICME, see Box 2.) The activities are geared to action. Dr. Ward postulated that the greatest challenges facing the government agencies in implementing the MGI are the coordination of effort and the understanding of lessons learned by different agencies.

Dr. Ward stated that open access to research results through open access publishing is growing, and it is creating an entirely new model for publishing. The government is encouraging open access via data management plans required in proposals—both the National Institutes of Health (NIH) and NSF have such requirements—as well as through a White House directive on public access (see Dr. Stebbins's presentation summary, below, for more information). Some research disciplines, such as crystallography and evolutionary biology, have taken the initiative by adopting their own data archiving policies.

Dr. Ward concluded with the following summary statements:

- Materials data have intrinsic value that can enhance the research and development process.
- Several technical and cultural factors make the capture, archiving, and sharing of materials data difficult.

BOX 2
Integrated Computational Materials Engineering

ICME is "the integration of materials information, captured in computational tools, with engineering product performance analysis and manufacturing-process simulation" (NRC, *Integrated Computational Materials Engineering*, p. 9). It is a process by which materials, manufacturing processes, and component design can be optimized long before the components are fabricated (see Figure B-2). It is considered a promising emerging discipline that is still under development.

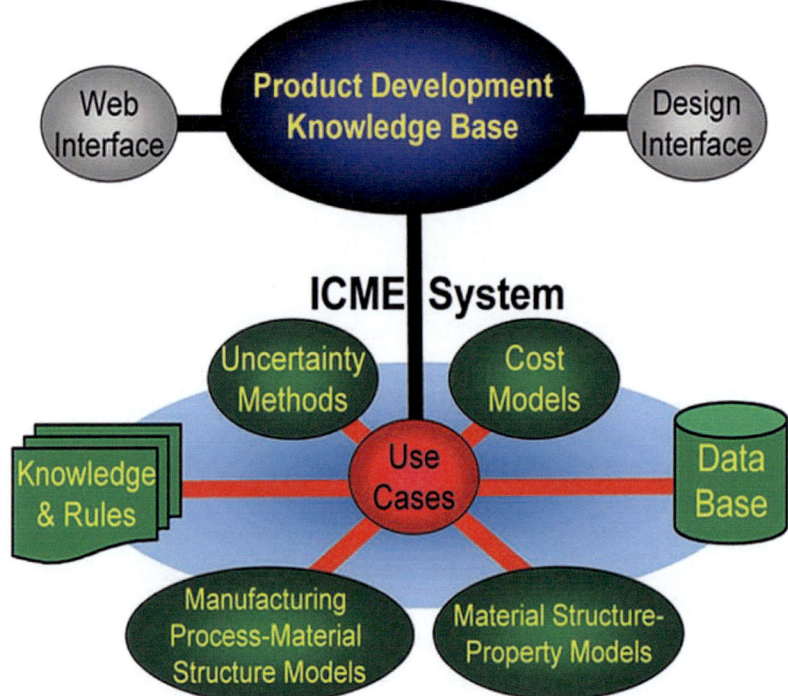

FIGURE B-2 An ICME system unifies materials information into a holistic system that is linked by means of a software integration tool to a designer knowledge base containing tools and models from other engineering disciplines.

SOURCE: NRC, *Integrated Computational Materials Engineering: A Transformational Discipline for Improved Competitiveness and National Security*. Washington, D.C.: The National Academies Press, 2008.

- The materials community appears willing to take on the challenge of materials data.
- Several MGI and community-led efforts are under way to guide materials data archiving.
- Broad discussion of materials data archiving within the community is gaining momentum and needs to be nurtured.

The question-and-answer period began with a discussion of data ownership. A participant asked about a scenario in which the data belong to whoever pays for it, not to the original data collector. Would this use of data without the knowledge of the persons who created the data not be dangerous? Dr. Ward pointed out that this is part of the normal course of science: One person creates data, another refines it. He gave several examples, such as data related to genomics, in which 500 people might access and analyze the data without having been part of its collection. He also pointed out that the field of evolutionary biology saw a 70 percent increase in data citations after setting up the Dryad program.

A participant asked if these challenges were being addressed by the materials genome community, or if they were merely being listed in the hope that another community would address them. Dr. Ward responded that the path forward is fairly clear, and that MGI hopes to follow the model of the evolutionary biology community in bringing the community together to decide what to do and how to add value.

Another participant pointed out that NSF and the NIH require data management plans in their proposals, and universities now have systems in place to publish data sets and provide them with a digital object identifier, which is citable forever. These might be models for the materials community. A data citation index might also be useful.

GE EFFORTS IN MATERIALS DATA: DEVELOPMENT OF THE ICME-NET

Russell Irving, Services Technology Leader, GE Global Research

Mr. Irving's talk drew on information from the Metals Affordable Initiative Workshop on Data Management for ICME in June 2011. At that meeting, he listed four key advancements in computer science that should be taken advantage of:

1. *Cloud computing.* This is elastic, use-as-you-go computing.
2. *Service-oriented architecture.* This allows for interoperability and sharing.
3. *Federated data.* This allows the owner of data to share some parts and

withhold other parts of databases. Users see data as if they were from one database.
4. *Business process management.* This is modern work flow software.

When combined, the four items can provide a collaborative ecosystem for materials development. The outcome at GE is the ICME-Net, which, according to Mr. Irving, fosters collaboration, enables the reuse of common analytics and processes, and builds knowledge accumulation and sharing. The interface to ICME-Net is a web browser providing access to all in GE who are cleared for access. Mr. Irving provided an example of ICME-Net for disk forging, shown in Figure 2.

Mr. Irving explained that GE had teamed with MIT to develop a collaborative ecosystem for open design and manufacturing; MIT was able to provide the software components needed, and together they built an environment for crowdsourcing military vehicle design. While the project did not move to Phase II, they were able to develop a robust prototype. GE leveraged the GE/MIT collaborative

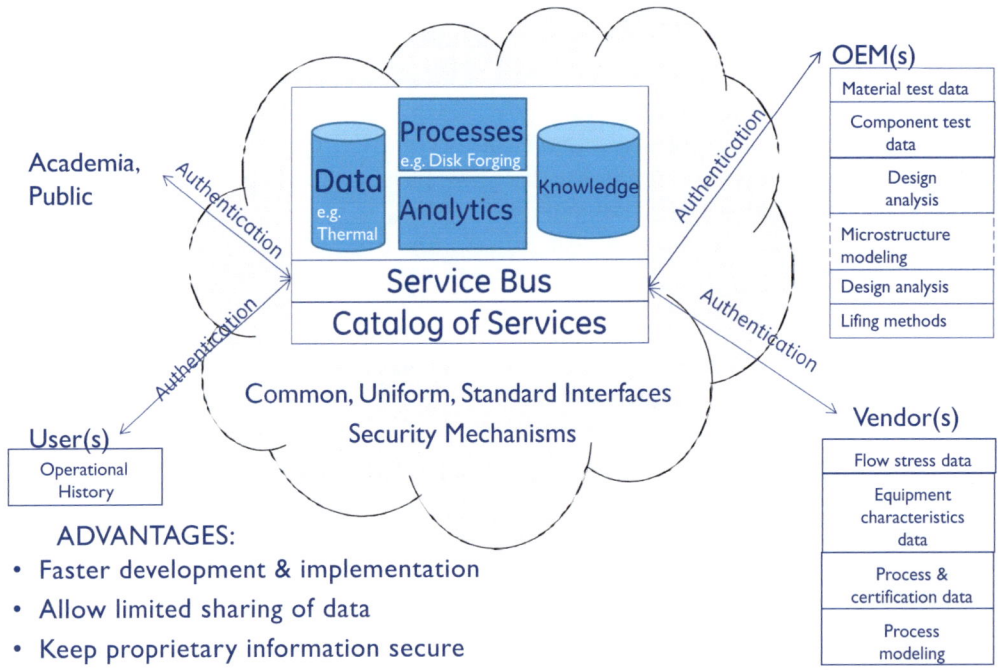

FIGURE 2 ICME-Net for a disk forging use case. SOURCE: Russell Irving, GE Global Research, presentation to the Metals Affordable Initiative Workshop on Data Management for Integrated Computational Materials Engineering, Slide 6.

ecosystem platform technology to develop an innovation infrastructure for ICME-Net. The initial vision for ICME-Net consisted of the following goals:

- Enable geographically distributed collaboration on development, testing, and demonstration of ICME technology.
- Rapidly disseminate technology development through a "marketplace" of materials engineering models and processes.
- Enable the construction of large, complex simulations for ICME.
- Attract and sustain an ICME community.
- Provide an opportunity for both open source and retention of intellectual property access to technology and best practices.

Mr. Irving then described ICME-Net in more detail. The goal is to grow a collaborative ecosystem. ICME-Net intends to build a marketplace that will subtly motivate the community to participate. This will enhance productivity, reduce development cost and time, curate model development, and make businesses more responsive. Projects consist of both components and services, and users can also be contributors.

Mr. Irving then discussed three use cases for ICME-Net. The first was the ceramic matrix composites project. ICME-Net provides a single location for the storage of experimental data and analysis. Users can decide whether to store a calculated result. The single location provides an auditable trail of every activity; it can be searched to find analysis done on a particular day or by a particular person. Mr. Irving said the interface has some similarities to Facebook, in that users can say whether they "like" a particular service. In the marketplace, therefore, the best services will accumulate the most "likes." Mr. Irving explained that each element (users, concepts, components, subcomponents, and services) is kept small. Curation and metadata management can be incorporated as the project develops.

Next, Mr. Irving described the Materials Applications Engineer for rotors, the second use case. This project results in large data sets, consisting of cut-up and microstructure data, as well as results from forging and heat-treatment models.

The final use case described was in alloy development. Users build small pieces of software to open the files and analyze the data one section at a time. The ICME-Net application contains a graphical user interface to lay out work flows and show linkages. The execution of the analysis is conducted in the cloud, and the curation and metadata management are separated from the data themselves.

Mr. Irving noted that ICME-Net can establish basic services for a particular application within several weeks. It was not originally intended to be used as a productivity tool, though that is now a top benefit. More functionality will continue to be added. He concluded with three main ideas:

1. The ICME concept is not new, and there are similar needs in other application areas, such as cyberinfrastructure.
2. Additional computing and software can extend the ICME-Net capability to, for example, a supply chain scenario.
3. There remain many outstanding issues associated with intellectual property, export control, proper business models, and other policy concerns.

In the discussion session, someone asked who the users of ICME-Net are. Mr. Irving responded that currently the users are all internal to GE, although GE is working to allow vendor access as well. The users are materials scientists with data that need to be analyzed.

SMART MANUFACTURING: ENTERPRISE RIGHT TIME, NETWORKED DATA, INFORMATION, AND ACTION

Jim Davis, Vice Provost, Information Technology, and Chief Technology Officer, University of California Los Angeles

Dr. Davis pointed out that his talk would not focus on "big data" per se. He said his focus is manufacturing—but, as it happens, much is data-oriented in manufacturing right now. He began by defining some of the terms in the title of his talk. He explained that he means for the term "enterprise" to have as broad a scope as necessary, whether that means from factory to supply chain or across units. "Right time" is similar to real time, but it is associated with the window in time for taking action and not the rate of data collection. He stressed that "action" is an important element of smart manufacturing.

Dr. Davis then described the Smart Manufacturing Leadership Coalition (SMLC), which he cofounded in 2006. SMLC uses an industry-driven strategy to make more information available to the manufacturing community. Its current focus is implementation. He explained that the manufacturing community generally wants to share services (or "apps"), not proprietary data. He said that in his role as UCLA's chief technology officer, he interfaces with the manufacturing operations community, where he sees an emphasis on smart manufacturing, cyberphysical systems, and the Internet of Things (IoT).[6] In his role as head of information technology for UCLA, Dr. Davis interfaces with the information technology community, where he sees a focus on enterprise resource planning, big data, cloud computing, and mobility.

---

[6] IoT is the network of uniquely identifiable physical objects (such as sensors or actuators) embedded throughout a network structure.

Dr. Davis then provided several different examples of smart manufacturing systems and the role of data in those systems:

- *Example 1: Smart Manufacturing at General Mills.* This is an example of network-based manufacturing. Dr. Davis said that a "material" is a statistical composite of its constituent elements. In this example of General Mills, one of the "materials" is Cheerios. The material for Cheerios is supplied by farms, all of which are subject to varying conditions such as weather and water accessibility. It would be helpful for General Mills to have basic data from each of the many farms in its supply chain, such as lot size and other information about the lot. There are manufacturing constraints as well, as the production facility must comply with cleanliness and contamination standards, and the Food and Drug Administration has tracking and traceability requirements. Dr. Davis explained that General Mills has a "green light" procedure, in which the constituent elements are analyzed for readiness to be put into production; the recipe is confirmed; the material is made; and the product is released upon confirmation that it meets requirements. Dr. Davis noted that while the Cheerios application may not qualify as big data, a lot of supporting data are involved in the manufacture of a Cheerio.
- *Example 2: Smart Manufacturing at Praxair.* Praxair supports oil and gas production at 40 or so facilities worldwide. Its furnaces must be kept in a certain temperature range. Dr. Davis explained that Praxair is currently very conservative about that range, which leads to waste heat and extra associated expense. Praxair would like to manage its risk differently to realize the potential for significant cost savings. The furnaces cannot have sensors in them, as the environment is too harsh. Instead, Praxair has turned to high-fidelity modeling in production. This project involves data at many different scales.
- *Example 3: Smart Manufacturing at General Dynamics.* At its Scranton, Pennsylvania, plant, General Dynamics has activities in heating and forging as well as cutting and machining. It is attempting to match the requirements for heating and forging to the output of the cutting and machining operations to better manage its energy usage and production efficiency.
- *Example 4: ICME at Caltech's Materials and Process Simulation Center.* This center assesses the manufacturing interface with materials models. The center is often asked about material risk and response and would like to support an operational mode, such as ICME. The center is looking at infrastructure to allow the models to move seamlessly into manufacturing. It is also working with Caltech/JPL on integrated systems design, which

looks at how design models can inform the system and how to use models in production.

Dr. Davis said that smart manufacturing is based on a testbed approach. He explained that the SMLC's portfolio of problems includes the following: smart machine operations, high-fidelity modeling, dynamic decisions, enterprise and supply chain decisions, and design and planning. All of these elements are data-intensive in some way, and an infrastructure is needed to manage these data appropriately.

Dr. Davis described smart manufacturing as a multilayered system, and he postulated that improvements can typically be made at points of handoff, or seams. Seams can be between different departments or vendors, between designers and manufacturers, or between business systems and control/automation. At the lowest layer, the microlayer, the focus is on insertion, rapid qualification, ICME, and informing control systems. There is a short time constant associated with these functions. The next layer, the mesolayer, is a much larger space and consists of the operational decisions. The focus is on operational performance and goals, maintenance, dynamic trade-offs, and people. The upper layer, the macrolayer, focuses on supply chain information and transitions to outside the company. Dr. Davis pointed out that there are seams within each layer as well as seams across layers. The time constants are different across the different layers, which creates seams as well. It is important to orchestrate applications—that is, manage the work flow. This construct allows one to manage time (a "window of action"). The work flow can be analyzed and then used to generate projections about the output. It also allows for tracking and traceability.

Dr. Davis then defined smart manufacturing intelligence and work flow. Smart manufacturing intelligence is characterized by

- Applications that can share data, data that can share applications, and applications that can connect to applications to achieve horizontal enterprise views and actions.
- Orchestration of standardized decision work flows based on structured adaptation and autonomy.
- Actionable data, trust, and visibility across the supply chain.
- In-time, in-production qualification of materials, products, and actions.
- In-time, in-production, multidimensional (business, operations, supply chain, customer, maintenance, energy) performance and adaptation.
- Cross-company operational data to improve performance.
- Evolvable design models in manufacturing.

The SMLC also defined work flow, stating that the smart manufacturing work flow enables a dynamic orchestration of manufacturing steps across different time constants and seams, including the supply chain, without losing control of state. It is hosted on an interoperable, accessible, affordable, secure, and reliable hybrid cloud platform and supports commercial products and services, research and development needs, and academic interests.

Dr. Davis explained how smart manufacturing interfaces with the ISA-95 standard.[7] He said that the SMLC takes advantage of standards but has not itself been involved in standards development. With respect to the definitions of ISA-95 Levels 0, 1, 2, or 4, the coalition is most interested in ISA-95 Level 3, which relates to manufacturing operations management. Level 3 seems to be a growing space with the most interfaces and thus offers the greatest opportunity for improvements in efficiency.

The SMLC looked at the main reasons the coalition stays together: Most SMLC members are interested in issues that extend beyond any one company. These issues are, for example, related to risk and organizational constraints. Data are part of each element.

The SMLC used a market-driven approach to identify its work flow, focusing on work flow as a service. The SMLC used the smartphone model of apps, toolkits, and so on. "App" is used here very broadly as a general-purpose term that allows systems to interface. The app model allows for enhanced flexibility. A work flow schematic is shown in Figure 3.

Dr. Davis explained that the SMLC uses a market-driven approach. Mapping and interface apps retrieve and manipulate data and map context. The apps should be put somewhere visible, and users should be able to understand how the apps are to be used and how well they perform. The SMLC is also interested in infrastructure that allows for the contribution of apps, broadening the space of innovation. The market is also driving standards-setting. Dr. Davis said this is equivalent to assembling a stack where the lower levels focus on the IoT. The SMLC focuses more on the middle layers, related to smart systems. Big data is the focus of the top of the stack; the amount of data increases as one rises through the stack. Currently, no one wants to share data, but there is a willingness to share apps and their use for different data applications. Dr. Davis pointed out that data valuation is a critical issue; when benefit is derived from the data, understanding the intellectual property becomes important, as does developing a business model to capitalize on that benefit.

Dr. Davis explained that there are many parallels in the health care system. The health care system also has value associated with patient data, as well as derivative intellectual property. One can think of companies as patients.

Dr. Davis noted that the app terminology relates smart manufacturing to

---

[7] ISA-95 is an international standard for the integration of control systems for manufacturing and processing operations. See http://www.isa-95.com for more information. Accessed February 25, 2014.

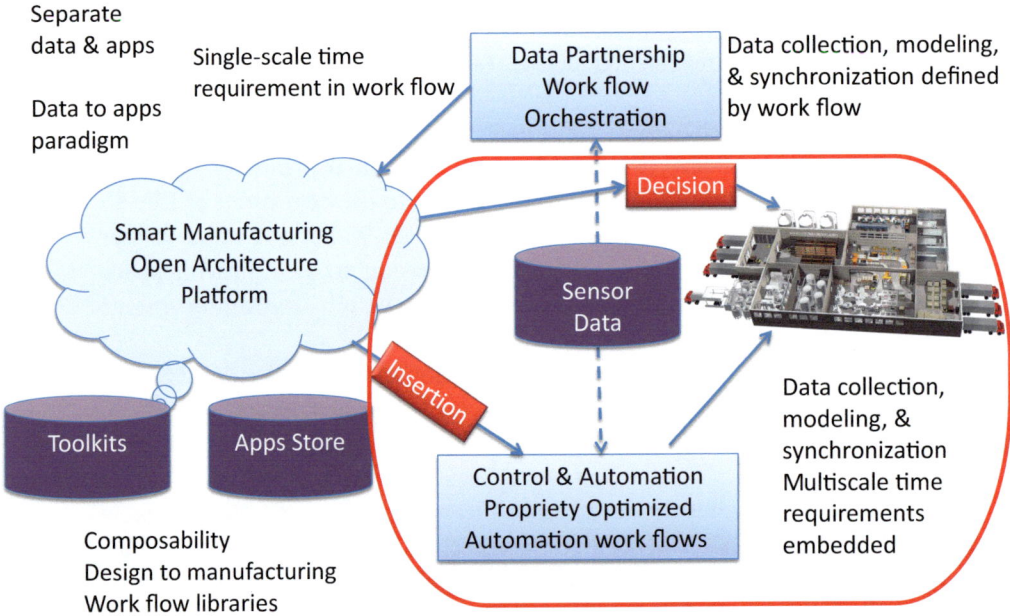

FIGURE 3 Smart manufacturing work flow. The sections circled in red at the bottom right are related to infrastructure. SOURCE: Jim Davis, University of California, Los Angeles, presentation to the committee on February 5, 2014, Slide 15. Courtesy of Smart Manufacturing Leadership Coalition.

smartphones. The chip layer in a smartphone is analogous to the SMLC work-flow-as-a-service layer; carriers match up to large manufacturers because both groups address issues related to matching platforms and compatibility. Also, smartphones have free core apps as well as paid apps, just as in the smart manufacturing realm. One distinction is that the smart manufacturing community is linking the data flow between existing apps. The apps themselves are not new, but the toolkits (work flows assembled to have a certain function) are the most valuable to the manufacturing community.

Dr. Davis concluded by saying that smart manufacturing cannot be addressed piecemeal. The coalition remains cohesive in its attempt to take a comprehensive view of how to proceed in smart manufacturing, focusing on the architecture to enable and orchestrate the apps while allowing the marketplace to decide the use.

In the discussion period, a participant pointed out that smart manufacturing is customer-driven. In the defense community, the notion of a customer may be less clear. The military is a consumer, but not necessarily a customer. Dr. Davis responded that whereas the focus would have to be less centered on the consumer when applying this to the DOD, the notion of orchestrating and coordinating is still important. Another participant suggested it may be better to consider the defense industrial base as the customer.

Dr. McGrath made some specific points related to Kepler.[8] He pointed out that the smart manufacturing work flow is based on Kepler, which is a scientific platform that has been used for many years for scientific work flows. He pointed out that Kepler currently seemed to be working well, but as the system becomes vendor-agnostic and moves into the open source, the specific platform may need to be reassessed. He pointed out that Kepler will need to be able to capture metadata.

Dr. Schafrik asked about the differences between small manufacturers and larger ones. Dr. Davis said that the smaller manufacturers do not have the time or money to address the issues related to smart manufacturing. If a smaller manufacturer is given the tools to manage information for multiple customers, it is likely to use them.

DISCUSSION

A participant asked about access to DOD data. The response was that AFOSR, ONR, DARPA, and others are exploring ways to share data generated under a DOD contract with the broader community.

Another participant asked about validating data, particularly when using advanced manufacturing techniques. Dr. Irving said that there is more emphasis right now on improving advanced manufacturing techniques; once the technique is refined, one can consider which process parameters to capture.

A participant argued that, for the small-scale researcher, data collection is outpacing computing capabilities. In the case of neutron beams, the bottleneck is processing the data rather than gaining access to beam lines. Continued progress is needed to provide new data analysis programs.

A participant pointed out that an industrial manufacturer such as General Mills has commercial supply chains. However, DOD does not. It can be difficult to keep all points of the DOD supply chain engaged when demand falls off. Another participant noted that an older NRC report commented on the dual-use industrial base (National Research Council, 1999). That report pointed out that, in a downturn, a company that is optimized for the defense industry cannot be readily commercially successful at the same time. There are no technical constraints, but the business models do not match well.

---

[8] Kepler is a free, open source, scientific work flow tool. See http://kepler-project.org for more information. Accessed February 25, 2014.

A participant concluded the discussion period by noting that the materials community has needs in all four V's of big data (volume, velocity, variety, and veracity) but at different levels of urgency. The materials community may not be as limited by computational technology as other fields are, though there are still unresolved issues in how to manage and analyze data.

Session 3: Big Data Issues in Manufacturing

Session 3 of the workshop focused on the application of big data concepts in manufacturing. Adele Ratcliff, Office of the Secretary of Defense, had originally been scheduled to present during Session 3 but was unable to participate. Presentations were made by Jesse Margiotta, DARPA, and Wayne Ziegler, Army Research Laboratory.

DATA NEEDS TO SUPPORT ICME DEVELOPMENT IN DARPA OPEN MANUFACTURING

Jesse Margiotta, Technical Advisor, DARPA

Dr. Margiotta began by saying that today's qualification and certification paradigm for parts and processes is fraught with difficulties. The methods are empirical, sequential, and iterative, leading to potentially prohibitive increases in cost and time. He said that the greatest challenge in qualification and certification, however, is its uncertainty. Dr. Margiotta pointed out that uncertainty in the qualification and certification process adds risk to a project, preventing new technologies from being incorporated into larger systems. As a result, the current qualification and certification paradigm creates a barrier to technology innovation and adoption. To counteract this, DARPA has begun the Open Manufacturing initiative; its main goal is to build and demonstrate a rapid qualification framework that aims to comprehensively capture, analyze, and control manufacturing variability. Dr. Margiotta explained the guiding principles for DARPA's Open Manufacturing program:

- Identify critical parameters, variation, and limits early in the process.
- Reduce testing and development iterations.
- Predict location-specific probabilistic performance.
- Build confidence in new technologies or qualification processes.
- Accelerate process maturity and systematic process reassessment.

Dr. Margiotta then described a project DARPA is developing with Honeywell Aerospace and several other team members. The project aims to develop rapid qualification of powder bed fusion additive manufacturing processes—in particular, direct metal laser sintering (DMLS). The general approach consists of the following elements: parameterize the manufacturing process; implement new sensors into the manufacturing process; incorporate an ICME construct that links process to materials to properties; and apply rigorous model verification and validation to understand the confidence limits. In this way, process parameters are linked to quantified, location-specific properties of the as-manufactured part. Dr. Margiotta showed a schematic of the critical elements for rapid qualification (Figure 4). The constituents to enable rapid qualification are shown in blue at the top of the figure. The supporting elements are shown below that; many of them—such as sensing, linking sensing capability to quality assurance, and microstructure property models—still need to be developed. Dr. Margiotta pointed out that the business cases and implementation plan are particularly important, as they will affect the usage and acceptance by the broader community. He stated that the architecture consists of increasing layers of complexity, including difficulties with the interfaces between different elements of the system.

Dr. Margiotta then explained the informatics associated with the additive manufacturing process. First, experiments are conducted to define the processing window, which is then refined through additional experiments to determine the optimal site within that window. This leads to a semioptimized process and overall improved material properties. The energy input density can also be measured and correlated with the quality of the consolidated material. In addition, the build chamber is instrumented to provide real-time monitoring of process parameters. The sensors have been able to capture a large quantity of high-fidelity data; at this point, about 1 TB of sensor data are collected for each DMLS build.

Dr. Margiotta then moved to the ICME construct, which uses process–microstructure–performance models to simulate the manufacturing process. Dr. Margiotta explained that the current simulation takes several days to a week to complete, which is much too long. These tools need to be further developed and simplified. The ICME construct consists of the following elements:

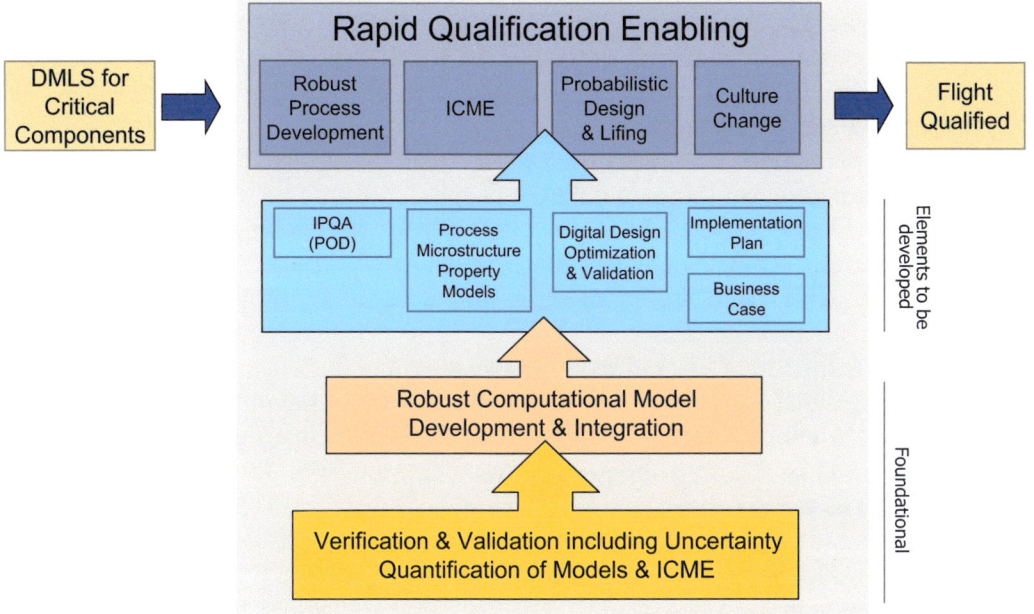

FIGURE 4 Critical elements of a rapid qualification system. SOURCE: J. Margiotta, Defense Advanced Research Projects Agency, presentation to the committee on February 6, 2014, Slide 4.

- Computationally intensive, physics-based models to simulate the manufacturing process. These models simulate the laser interaction with the powder bed, including thermal profiles and heating rates.
- Microstructural models to predict stresses, grain size, strain hardening, and other variables.
- Yield strength prediction tool.
- Uncertainty quantification to understand the relationship between processing and properties and the sensitivity of this relationship.

Dr. Margiotta said that the framework is the most critical element of the system. Once the general framework is in place, tools can be swapped in as they are developed. He noted that the tools are still under development and that much work remains.

Dr. Margiotta explained that the Open Manufacturing project was one of the first to extend verification and validation and uncertainty quantification to ICME processes. The Open Manufacturing project intends to draw on the work and standard practices developed in other fields in which simulation is well developed and rigorously validated. These same methods can be transitioned into the materials and manufacturing arena.

A participant asked what is meant by "rapid" in this context, because the term can mean different things to different people. Dr. Margiotta responded that a typical qualification effort takes at least several years, with a long development effort prior to that before the qualification is begun. He was hesitant to associate a specific number with "rapid" but pointed out that there are significant time savings to be had in qualification efforts.

A participant also asked about the meaning of probabilistic design. Dr. Margiotta explained that one identifies the worst defects possible and puts them into the most critical locations, then, using the minimum material properties, designs the part so that it will be able to withstand that worst-case scenario. Alternatively, probability should be used to understand the likelihood of the defect, its location, and its effect and should then optimize the design accordingly.

Dr. McGrath asked about the data that have resulted from the Honeywell additive manufacturing project. Dr. Margiotta responded that the intent is for these programs to provide the data in a data archival tool, as is described in Mr. Ziegler's talk (below). Access will be provided to other government agencies, with the details on broader distribution still to be determined. The process and materials data will not become proprietary. Dr. Margiotta was asked if this constituted a "big data" problem. He responded that the Honeywell project is generating considerable amounts of data, but it is not considered big data based on today's definition. However, the materials manufacturing community does not currently have the ability to manage and analyze even this relatively modest amount of data.

Dr. Margiotta also pointed out that DARPA, along with ARL and other program partners, is developing methods to standardize data fields and metadata fields for materials and materials processing.

THE MATERIALS INFORMATION SYSTEM

Wayne Ziegler, Materials Engineer, Army Research Laboratory

Mr. Ziegler began his presentation by explaining the value of a materials information system. He noted that, while it is not the case for this audience, he often needs to convince his listeners of the value of the materials information system approach; people often are more concerned with intellectual property or do not understand the problems that currently exist. He pointed out that a materials information system does the following:

- Enables researchers to work more quickly and intelligently.
- Reduces duplication of effort in test and evaluation, which correspondingly reduces costs.
- Stops data loss and ensures data are available for the next generation.
- Improves data consistency and quality.
- Improves work processes and throughput.
- Accelerates implementation.

Mr. Ziegler said that a successful data management plan uses a systems engineering approach and includes four main components: capture, analyze, deploy, and maintain. He suggested that DOD had historically had difficulty with maintaining programs that manage material and process data due to the challenges of a mobile workforce and shifting budget considerations. NASA also has a long history with strict materials data management, as several major catastrophes at NASA were related to materials issues. As a result, ARL is working with NASA to identify lessons learned and leverage NASA's experience and IT infrastructure resources. Current challenges associated with the development of materials data management plans, both in industry and DOD, include these:

- Lack of direction.
- Lack of adequate resources.
- Lack of a return-on-investment business case.
- Lack of agreement. Not all companies or agencies believe that all data should be shared, and the cultural mindset needs to be changed.

Mr. Ziegler said that the goal is to build a DOD resource for materials and process information. The DOD resource, Materials Selection and Analysis Tool (MSAT), is currently hosted by NASA as an independent component of the NASA Materials and Processes Technical Information System (MAPTIS) system. He went on to say that the MSAT program has a strong partnership with DARPA and its Open Manufacturing program (see Dr. Margiotta's presentation, above).

Figure 5 shows a vision of a materials information system. It begins with experimental methods; Mr. Ziegler pointed out that we tend to lose metadata in this area, and experimental methods and results should be part of an integrated database structure. Metadata include information related to testing conditions and program information necessary for data sets to be completely understood and, if necessary, validated through additional testing. Data mining techniques are then applied to the data, and the mined data are used to inform models. This process is iterative and requires the tracking of data pedigree, in other words, data about the data. Mr. Ziegler noted that as much as half of the data can be pedigree informa-

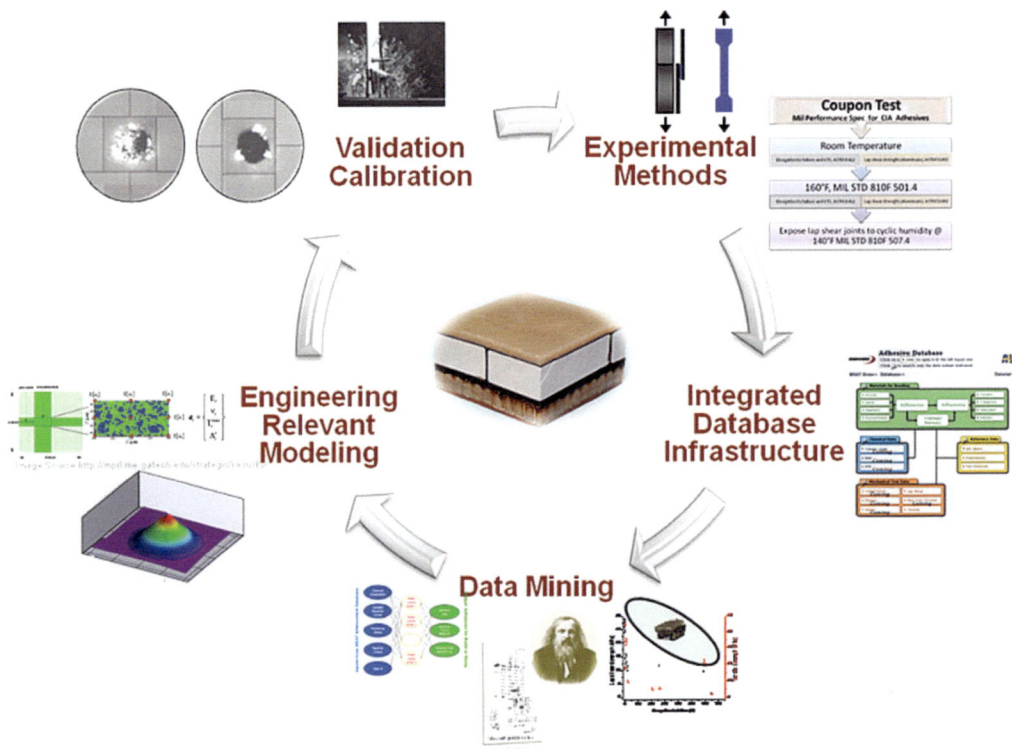

FIGURE 5 Materials information system. SOURCE: Wayne Ziegler, Army Research Laboratory, presentation to the committee on February 6, 2014, Slide 4.

tion. Relevant metadata are needed to compare data across data sets. Mr. Ziegler indicated that the Open Manufacturing project follows this general form.

Mr. Ziegler pointed out that MSAT is called a selection and analysis tool since its initial focus is making programmatic or research decisions based on the data sets available in a robust and timely way. MSAT has a wide approach to application, modeling, resource management, process approval, and improvement.

Mr. Ziegler then argued that there needs be a cultural shift in work flow management. The traditional work flow paradigm is to execute a task, collect and extract data, return a bigger data set, and pair it with separately recorded information about the process. However, when the collection of data is separated from the collection of process information, fidelity drops. Mr. Ziegler argued in favor of a new work flow paradigm, whereby when a process is executed, the data and the metadata are collected simultaneously. When data are extracted from this set, there would already be a link that coupled them in any future data processing. He went on to describe the steps in the materials information system:

1. *Define data sets.* Mr. Ziegler pointed out that any data that can be collected in a reasonable fashion should be collected, as it might prove useful later. ARL is still addressing this first step, and Mr. Ziegler explained that the data collection decisions are iterative; once they have started collecting a particular data set, researchers will likely determine that they will need other data as well.
2. *Define data management schema.* This process looks at how to organize and arrange the data. It is also an iterative process.
3. *Develop import templates.* Mr. Ziegler noted that ARL currently uses Excel; the templates are in a comma-separated values format so that they are software-agnostic. Several companies provide commercial data management packages or modeling software, and the objective is to build templates that can interface with a variety of commercial software.
4. *Use templates to import data.*
5. *Manage data.* This step includes defining access control and conducting verification and validation.
6. *Define the use cases.* The resulting information is used to define output templates.

Mr. Ziegler suggested that Steps 1, 5, and 6 are the most critical for the users. Once the use case is known, then the data can be exported in a useful way. Mr. Ziegler then discussed several technical considerations, including the following:

- Defining the main function of the system: Is it capturing a manufacturing process, or does it focus on identifying material properties?
- Integrating the system with existing systems and workplace practices with minimum impact.
- Understanding how system users will use the data.
- Data flow through the system. Mr. Ziegler noted that in the materials science and engineering technology area, this is not "big data" (yet), though it is on a large scale.
- The type of information to be handled.
- System setup, deployment, and maintenance.
- Responsibility for, and ownership of, the various system components. Mr. Ziegler said that this can be a contentious issue, as data have value. Defin-

ing data access may not always be technically challenging, but it can be a policy challenge.

Mr. Ziegler concluded by noting some practical considerations, including these:

- Not every user is an expert, so the user interface becomes critical.
- Materials and process data are usually incomplete.
- Data have value; access control is critical and potentially contentious.
- It takes time to rationalize and consolidate data. The better the system is at collection, the better it will be at consolidation.
- Data end users need data in diverse places and formats.
- Materials information systems need end users. Mr. Ziegler argued that there is no value in the system if it does not have end users. It can be challenging to identify and engage users.
- Designing a system from scratch is impractical.

A participant indicated that data ownership can be an obstacle to data sharing. He said that DOD contracts have many data requirements in them, and that aspect needs to be managed on the contractual side to ensure that the requirements are not cost prohibitive. He pointed out that there is a responsibility for sharing data among materials suppliers, original equipment manufacturers, and the government. In some cases, a supplier provides a material but no corresponding metadata. Agreements with suppliers can take 1-2 years to develop, which slows down innovation. Mr. Ziegler agreed that acquisition is an important element, though outside the scope of ARL's mission and activities.

Dr. McGrath asked for clarification on MSAT. Is it a tool for materials selection, with a correspondingly fairly limited user community? Or is it an element of a larger system within a larger community, with a framework surrounding it? What is the plan for scaling up past the Open Manufacturing project? Mr. Ziegler said that MSAT is both a materials selection tool and part of a larger system. MSAT's current focus is on where to store materials and processes and how to develop a clear interface with the modeling community.

The discussion then turned to standards. A participant stressed that the process for developing standard terminology is very difficult and slow. There is an ASTM committee for standards in this area. Companies do not like to fund their employees to do this type of activity, however, and the ASTM committee terminated its efforts because of insufficient community funding. Also, companies are not interested in attaching themselves to a certain format, as they are concerned they will be forced to share data. They prefer to keep information proprietary in their own formats. A few participants noted that the culture among researchers and companies is such that materials data are considered a competitive edge, and companies want to protect their intellectual property.

DISCUSSION

Valerie Browning, from ValTech Solutions LLC, opened the discussion by noting that the workshop speakers thus far have discussed materials challenges in variety and veracity, but not volume or velocity. In materials, therefore, it may be more important to think about information rather than big data. The materials area has an extra layer of extraction or analytics that is unique to this community. The questions then become, Who is responsible for developing the analytics? How can we manage work flow on different time scales? Who owns the analytics?

Another participant agreed that there is a data problem in materials science, but not a big data problem. He said that the data problem seems to center on data collection and the lack of sharing. He suggested that a mandate is necessary stating that any government-funded data must be put into a standard format, a step that is being considered by NSF and DOE. A DOD participant said that DOD has explored the idea of such a mandate but finds it time consuming and expensive and, in the end, concludes that the cost may outweigh the benefit. He indicated that the issue is more than one of data format; it includes questions about who owns and maintains data and where the information should reside. Someone else remarked that the NSF repository is not user-friendly. Another participant pointed out that it is fairly common for universities to have permanent storage facilities available and gave the Deep Blue program at the University of Michigan as an example.[1] However, other participants argued that these programs are expensive and do not always include metadata.

One participant believed that there is a data collection problem in manufacturing. Manufacturers need to organize data definitions, contextuals, and the meaning of operations and connect with ICME. Dr. Davis brought up the analogy of health care, and finds many parallels in the health care shift to digital patient records. Another participant noted that advanced manufacturing can be done on a small scale that is not necessarily part of a large corporation. This could add a layer of complexity, as a small player may not be interested in negotiating business plans.

---

[1] See http://deepblue.lib.umich.edu/ for more information. Accessed February 26, 2014.

Session 4: The Way Ahead

The final session of the workshop focused on the federal government's present and future projects and plans related to data in materials and manufacturing. Presentations were made in Session 4 by Julie Christodoulou, Office of Naval Research, and Michael Stebbins, Office of Science and Technology Policy.

LIGHTWEIGHT AND MODERN METALS MANUFACTURING INNOVATION INSTITUTE: IMPLICATIONS FOR MATERIALS, MANUFACTURING, AND DATA

Julie Christodoulou, Director, Naval Materials Division, Office of Naval Research

Dr. Christodoulou began her presentation by describing the Lightweight and Modern Metals Manufacturing Innovation Institute, known as the LM3I Institute (LM3II). The LM3II provides an opportunity to integrate materials into emerging manufacturing technologies in a meaningful way to ultimately produce better-performing systems and components for military and civilian applications, and to do so faster and more affordably. LM3II is part of a national network of institutes that focus on topics of national import and benefit. It will have many partners, and DOD is leading the effort. Dr. Christodoulou said that metals are uniquely dual-use. There is no conflict between military applications and commercial value; DOD welcomes commercial applications of materials technologies, because a broader marketplace will help reduce costs for the military. The goal of the LM3II is to demonstrate advanced manufacturing capabilities to enable systems that are lightweight, reliable, survivable, fuel efficient, affordable, and flexible while designing lightweight metals and applications that use them. Lightweighting has a near-term impact on fuel efficiency, flexibility, and overall costs for naval systems and platforms, and Dr. Christodoulou pointed out that many of these benefits were discussed in the National Research Council's report on lightweighting (NRC, 2012). She said that LM3II focuses on aluminum, titanium, magnesium, and other modern metals and on manufacturing readiness levels (MRLs) 4-7.[1] It is interested in incorporating ICME to shorten the design–production cycle. The goal is to have the program be self-sufficient in 5 years, with no dedicated funding after that time. Achieving this goal will require the identification of transition partners.

Dr. Christodoulou then briefly described the ICME process (see Box 2 in Dr. Ward's summary for more information about ICME). She argued that ICME is an important driver in the acceleration of development times.

Next, she briefly discussed DARPA's Accelerated Insertion of Materials (AIM) program. The AIM program was established to apply ICME to specific applications in the aircraft engine industry so as to reduce development time and costs. The AIM program focused on the designer's toolbox, including knowledge, heuristics, and collections of data sets. Dr. Christodoulou stressed the importance of being able to use that set of knowledge and heuristics to exert greater control over a material's structure and properties. She pointed out that ICME can help materials scientists communicate better with part designers, enabling them to take advantage of the complexity of a material's microstructure to design a better component.

Dr. Christodoulou then provided an example of a project with exceptionally complex data. The example related to grain growth in a titanium alloy. The project, led by Lauridsen and Voorhees, used high-fidelity three-dimensional synchrotron imaging techniques and phase field modeling to capture anisotropy effects. The project used successive images to study the progression of grain boundaries to understand the microstructure. While the data collection process took about a week, the data analysis took over 2 years because of the exceedingly complex data (Poulsen, Voorhees, and Lauridsen, 2013).

Dr. Christodoulou also stressed the importance of mixing models at multiple length scales to predict microstructural evolution and resultant localized properties. She pointed to the work of Li and Wang in phase field models of microstructure evolution. Phase field models developed at different length scales provide a useful

---

[1] MRL is a scale on which to assess manufacturing maturity and risk using a standard set of criteria. MRLs range from 0 (basic manufacturing implications identified) to 10 (full-rate production demonstrated). See http://www.dodmrl.com for more information about MRL definitions. Accessed February 27, 2014.

means to capture fundamental insights into phase transformation and deformation mechanisms and to establish physics-based models for engineering applications. Again, in order to do this, large and realistic volume elements must be explored. Materials naturally have a large amount of heterogeneity associated with them, and that space needs to be examined. The data need to be very rich in detail in order to be able to create an accurate model. These systems also require powerful analysis tools to acquire the data and develop models that are correspondingly rich in detail.

Dr. Christodoulou then discussed the Advanced Manufacturing Partnership and the MGI. The two initiatives were announced jointly in June 2011 and are synergistic. (See Box 1 in Dr. Ward's presentation, above, for more information about MGI.) Dr. Christodoulou pointed out that ICME is the primary tool to address the challenges presented in the MGI, and MGI is therefore the entity to enable widespread adoption of ICME—to make best practice become common practice. She also reminded the audience that there is a continuum in materials development: The phases of materials development (discovery, development, property optimization, systems design and integration, certification, manufacturing, and deployment) are all synergistic, and researchers in each phase need to be able to interact using a common language.

Dr. Christodoulou then discussed the National Network for Manufacturing Innovation (NNMI), a program that is designed to look at the interfaces among industry, academia, and government partners to identify gaps and provide the missing physical and intellectual middle ground. NNMI is a technology incubator for small organizations or multiagency groups. A schematic of the vision for NNMI is shown in Figure 6. Dr. Christodoulou described the NNMI as an incubator space, to which smaller or multicomponent groups can bring a problem or idea and test its viability. She noted that the NNMI seeks to provide opportunities for education at all levels, including sabbaticals, student co-ops, and internships.

Returning to the goals of the LM3II, Dr. Christodoulou emphasized the importance of bringing in industrial counterparts early in the design and development process; she noted that this is one of the central requirements for ICME. The transition pathways will then be established from the very onset of a project. She reminded the audience that LM3II is looking for self-sustainment—that is, independence from dedicated federal funding—in 5 years, so transition opportunities must be explored now. She believes that the LM3II will be formalized in the near future.

Dr. McGrath asked about the LM3II's expectations and whether it anticipates becoming like the open platform of Open Manufacturing. Dr. Christodoulou responded that LM3II would be a complex, public–private partnership for the next 5 years, with $70 million total investment from the government over that time and with expectations of more than that in matching funds. The institute is currently sorting through issues related to levels of engagement and intellectual property. She said that each project would likely have its own unique intellectual property

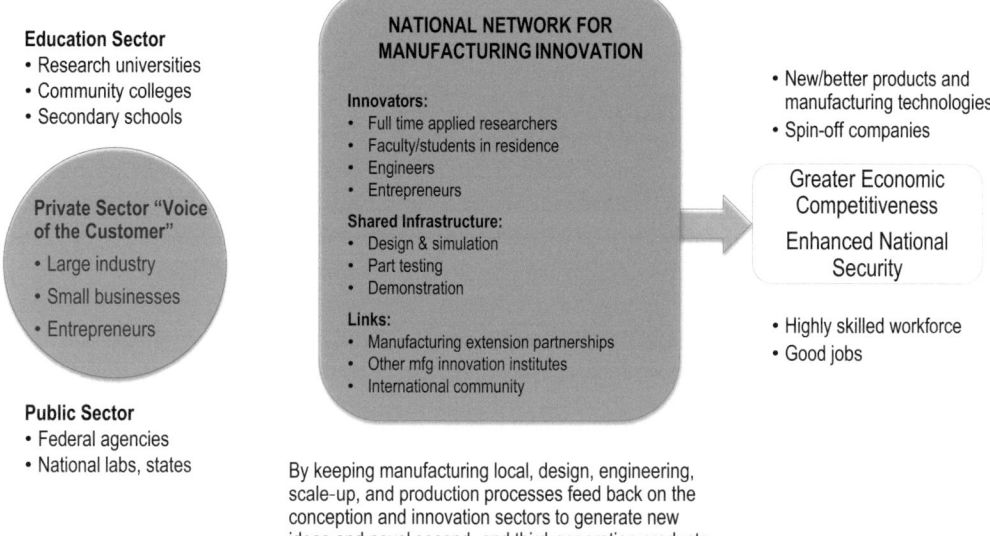

FIGURE 6 Elements of the NNMI. SOURCE: Julie Christodoulou, Office of Naval Research, presentation to the committee on February 6, 2014, Slide 28.

arrangements, although the approach to developing those arrangements can be established early. The LM3II is working toward common data formats while making sure to understand the different architectures under development.

Dr. Schafrik asked about the GE example, which consisted of a project in solid-state welding in titanium. He noted that there are unusual phases at the interface zone. He asked if those were under investigation in this project, or whether the project was considering only the monolithic material. Dr. Christodoulou responded that they hope to look broadly at manufacturing technologies, including joining.

A participant asked if Dr. Christodoulou has considered cartography to represent microstructure. Dr. Christodoulou said that some element of mapping has to be undertaken to represent microstructure. There is a need for standards to capture the properties. Another participant pointed out that the materials and manufacturing community borrows from the medical community in imaging. Dr. Christodoulou agreed and noted that the digital data set is not simply a digital image. Material properties, performance, and structure evolution must be quantified, along with any changes to those variables. Still another participant said that collections of microstructures tend to focus on small-scale changes but that it is important to simultaneously capture changes on longer scales.

DIRECTION OF POLICY

Michael Stebbins, Assistant Director for Biotechnology, Science Division, White House Office of Science and Technology Policy

Dr. Stebbins explained that he is the White House lead on openness and access to scientific data. The White House activity encompasses all fields of scientific study, and it has direct relevance to the materials community. He pointed out that many workshop participants noted that materials science does not have what is generally considered a "big data" problem." However, he would like for there to be one: He would prefer to see that the materials science community has too much data rather than not enough. Dr. Stebbins pointed out that, looking broadly across all scientific disciplines, it is difficult, if not impossible, to access the data that underpin figures in scientific journals. There are historical reasons for this, he argued, not practical limitations, as data storage is now inexpensive and widely available. The obstacles that prevent data access include cultural mindset, lack of suitable storage locations, lack of data standards, and insufficient funding for managing data sets. Dr. Stebbins pointed out that the biomedical research community has been approaching a crisis in that it has become virtually impossible to reproduce biomedical results. This is not because of widespread fraud, but because of a lack of access to protocols, underlying data, and process information. The biomedical community has developed good monotypic databases, with analytic tools superimposed on them. As a result, the community is able to share insights and accelerate development. Dr. Stebbins argued that genomics is a good example of a field in which data sharing is particularly critical. He added that, while monotypic databases are important and should continue to be used, something more may be necessary for a large corpus of data and said that a 3-4 PB per year database is a good prototype starting size. The White House is in the process of working with scientific journals to require that any article published in them have referenced data with a persistent identifier, including metadata and proper credit.

Dr. Stebbins pointed out that the only "currency" available today to a research scientist is the scientific publication in a journal. It may be possible to add a new currency: the actual data. He said that frequently a single paper may have many coauthors, and the individual contribution of each author is unclear. This is particularly true in certain fields, such as high-energy physics. Published data may be able to clarify the individual's role. In some fields, publishing the data may not be feasible or useful; in others, such as biology, it will dramatically accelerate development and allow for reproducibility. He asked the workshop participants if materials science would find the publication of data useful. Any potential advances, unintended consequences, or possible hazards in materials science should be explored now, before engaging the relevant publishers.

Dr. Stebbins pointed out that the government has had relatively little impact on increasing data sharing. By contrast, federal laws tend to focus on preventing the government from forcing the sharing of results. The Bayh-Dole Act,[2] for instance, encourages researchers to retain their discoveries for the purpose of commercialization. However, the government has not had much experience in the promotion of data sharing. The most successful instances of data sharing have been driven by publishers, who seek to ensure data quality in their publications.

A participant pointed out that the National Nanotechnology Initiative (NNI) has a focus area in informatics and is working with relevant journals on data access as well. Dr. Stebbins said that it would be important for conversations to be coordinated with White House policy and with the bigger picture surrounding information sharing.

Another participant pointed out that universities are struggling with open publishing vs. licensed publishing. Universities must pay to access journals. These costs may be amplified if the access suddenly includes additional costs for data access as well. Dr. Stebbins said that the initial pilot programs OSTP is considering will be free. The researchers (not the publishers) will deposit their data with a registered third party. In time, standards will develop for the database, and publishers may require a specific data management solution that costs additional money. As an example, the journal *Nature* has an agreement with the company FigShare. Participants worried that this additional cost for having FigShare manage the data is likely to be passed on to the authors, and publishing in *Nature* is already quite expensive. A high-end journal such as *Nature* is likely to succeed with this model, but other journals may find that additional data costs will discourage article submissions. Dr. Stebbins pointed out that other private sector entities will likely become more engaged in data storage and management, including the nonprofit community. A participant asked who would select the data storage site. Dr. Stebbins replied that the journal could say that it has a preferred database or a partnership with a particular data-sharing site. Alternatively, the journal could establish the data standard. Data longevity will need to be part of the standard; longevity would need to be guaranteed. Self-storage is unlikely to meet any longevity standard.

Various participants discussed questions about access to data that still needed to be addressed, which data set(s) needed to be provided, and data standards. Someone argued that it would not be sufficient to upload a data set; authors must upload models as well. Dr. Stebbins agreed that models should be part of the metadata, including specific algorithms and software versions. Someone else pointed out that publishers are a good point of leverage in the academic community. In

---

[2] The Bayh-Dole Act of 1980 amended patent and trademark law to permit universities and small businesses to retain intellectual property rights to work conducted using federal funding (Public Law 96-517).

communities such as pharmaceuticals and other industries, however, publication is not a high priority. The data in those instances are used to build business and are considered proprietary. It becomes difficult, if not impossible, for the outside community to gauge the quality of those data. The participant supposed that it would be very challenging to try to move into the commercial space with this proposal.

Wayne Ziegler and Chuck Ward both said that this open data initiative is consistent with and complementary to the activities in ARL and AFRL. Dr. Ward said that both their initiatives have been in communication with NIST and the relevant journals. He said that they are working to develop standards for CALPHAD data. They will next work with issues related to more complex data. He noted that it is fairly easy to come to a set of data standards for crystallography and CALPHAD, but other more complex data sets, such as three-dimensional data, will be more difficult.

A participant voiced concern that materials research has a lot of associated information or metadata, including sample preparation, and tracking the amount of metadata associated with an individual sample may be onerous. Dr. Stebbins pointed out that while this could be true it was still his belief that it is tractable.

References

NRC (National Research Council). 2013. *Frontiers in Massive Data Analysis*. Washington, D.C.: The National Academies Press.

NRC. 2012. *Application of Lightweighting Technology to Military Vehicles, Vessels, and Aircraft*. Washington, D.C.: The National Academies Press.

NRC. 1999. *Defense Manufacturing in 2010 and Beyond: Meeting the Changing Needs of National Defense*. Washington, D.C.: National Academy Press.

NRC. 1995. *Preserving Scientific Data on Our Physical Universe: A New Strategy for Archiving the Nation's Scientific Information Resources*. Washington, D.C.: National Academy Press.

Poulsen, S.O., P.W. Voorhees, and E.M. Lauridsen. 2013. Three-dimensional simulations of microstructural evolution in polycrystalline dual-phase materials with constant volume fractions. *Acta Materialia* 61(4): 1220-1228.

Appendixes

A

Workshop Statement of Task

An ad hoc committee will convene a series of three 2-day public workshops to discuss issues in defense materials, manufacturing and infrastructure including: 1) Globalization of Defense Materials and Manufacturing; 2) Big Data in Materials Research and Development; and 3) Materials State Awareness. The committee will develop the agendas for the workshops, select and invite speakers and discussants, and moderate the discussions. The workshops will use a mix of individual presentations, panels, and question-and-answer sessions to develop an understanding of the relevant issues. The workshop topics will highlight some recent developments in the fields. Key stakeholders will be identified and invited to participate. Individually authored workshop summaries will be prepared separately by a designated rapporteur after each workshop in this series.

B

Workshop Participants

Bharat Bhushan, Ohio State University
Valerie Browning, ValTech Solutions, LLC
Dennis Chamot, National Research Council (retired)
Julie Christodoulou, Office of Naval Research
Dan Crichton, NASA Jet Propulsion Laboratory
Jim Davis, University of California Los Angeles
Jesus de la Garza, Virginia Tech
Robert Dowding, U.S. Army Research Laboratory
Barry Farmer, U.S. Air Force Research Laboratory
David Forrest, Department of Energy
Steven Freiman, Consultant
Lisa Friedersdorf, National Nanotechnology Coordination Office
Rosario Gerhardt, Georgia Institute of Technology
Natalie Gluck, Institute for Defense Analyses
Ken Hix, GE Aviation
Russell Irving, GE Global Research
Robert Latiff, R. Latiff Associates
Alexis Lewis, Naval Research Laboratory
Steven Linder, Office of the Secretary of Defense
Kenny Lipkowitz, Office of Naval Research
Jesse Margiotta, Defense Advanced Research Projects Agency
Thom Mason, Oak Ridge National Laboratory
Suveen Mathaudhu, North Carolina State University

Peter Matic, Naval Research Laboratory
Anne Matsuura, Optical Society of America
Michael McGrath, Analytic Services, Inc.
Heather Meeks, Defense Threat Reduction Agency
Jed Pitera, IBM
Ward Plummer, Louisiana State University
Siddiq Qidwai, Naval Research Laboratory
Mike Rigdon, Consultant
John Rumble, R&R Data Services
Robert Schafrik, GE Aviation
Dave Shepherd, Department of Homeland Security
Lewis Sloter, Office of the Assistant Secretary of Defense for Research and Engineering
Michael Stebbins, White House Office of Science and Technology Policy
David Stepp, U.S. Army Research Laboratory
Denise Swink, Private Consultant
Galip Ulsoy, University of Michigan
Hayden Wadley, University of Virginia
Chuck Ward, U.S. Air Force Research Laboratory
Scott Weidman, National Research Council
Jennifer Wolk, Naval Surface Warfare Center
Jeff Zabinski, U.S. Army Research Laboratory
Wayne Ziegler, U.S. Army Research Laboratory

C

Workshop Agenda

FEBRUARY 5, 2014

8:30 am	Welcome and Meeting Objectives	Michael McGrath, Chair, and Robert Schafrik, Vice Chair
9:00	**Introduction to Big Data**	
	Frontiers in Massive Data Analysis and Their Implementation	Daniel Crichton, JPL
9:40	IBM and Big Data	Jed Pitera, IBM
10:40	Biosecurity and Big Data	Dave Shepherd, DHS
11:20	Discussion	
1:00 pm	**Big Data Issues in Materials R&D**	
1:40	Physics in Big Data	Thom Mason, ORNL
2:20	MGI and Big Data Formats	Chuck Ward, Air Force

Appendix C

3:20	GE's Implementation of ICME for Materials Data	Rusty Irving, GE
4:00	The Smart Manufacturing Leadership Coalition and Big Data	Jim Davis, UCLA
	Discussion	

FEBRUARY 6, 2014

8:30 am	Welcome and What We Heard Yesterday	Mike McGrath, Chair
8:40	**Manufacturing Issues in Big Data**	
9:20	Open Manufacturing/Honeywell Data Analysis	Jesse Margiotta, DARPA
10:20	MSAT/MAPTIS Data Archiving and Process and Materials Data	Wayne Ziegler, ARL
	Discussion	
11:00	**The Way Ahead**	
11:30	The Lightweight Materials Institute	Julie Christodoulou, ONR
1:00 pm	The Direction of Policy	Michael Stebbins, OSTP
2:10	Discussion	
3:00	Closed Session—Planning	
	Workshop Adjourns	

D

Acronyms

ADARA	Accelerating Data Acquisition, Reduction, and Analysis
AFOSR	Air Force Office of Scientific Research
AFRL	Air Force Research Laboratory
AIM	Accelerated Insertion of Materials
ARL	Army Research Laboratory
CALPHAD	CALculation of PHAse Diagrams
DARPA	Defense Advanced Research Projects Agency
DHS	Department of Homeland Security
DMLS	direct metal laser sintering
DMMI	Defense Materials, Manufacturing, and Infrastructure
DOD	U.S. Department of Defense
DOE	U.S. Department of Energy
GE	General Electric
HIV	human immunodeficiency virus
ICME	Integrated Computational Materials for Engineering
IoT	Internet of Things
IPCC	Intergovernmental Panel on Climate Change

JPL	Jet Propulsion Laboratory	
LM3II	Lightweight and Modern Metals Manufacturing Innovation Institute	
MAPTIS	Materials and Processes Technical Information System	
MDF	Manufacturing Demonstration Facility	
MGI	Materials Genome Initiative	
MRL	manufacturing readiness level	
MSAT	Materials Selection and Analysis Tool	
NASA	National Aeronautics and Space Administration	
NIH	National Institutes of Health	
NIST	National Institute for Standards and Technology	
NNI	National Nanotechnology Initiative	
NNMI	National Network for Manufacturing Innovation	
NRC	National Research Council	
NSF	National Science Foundation	
ONR	Office of Naval Research	
OODT	Object Oriented Data Technology	
ORNL	Oak Ridge National Laboratory	
OSTP	Office of Science and Technology Policy (White House)	
SARS	severe acute respiratory syndrome	
SMLC	Smart Manufacturing Leadership Coalition	
TRL	technology readiness level	